R-3928-NSF

Multiplying Inequalities

The Effects of Race, Social Class, and Tracking on Opportunities to Learn Mathematics and Science

Jeannie Oakes
with Tor Ormseth, Robert Bell, Patricia Camp

July 1990

Supported by the
National Science Foundation

PREFACE

In its 1983 report to the nation, *Educating Americans for the Twenty-First Century,* the National Science Foundation (NSF) set an ambitious goal for precollege science and mathematics education: to provide "high standards of excellence for all students—wherever they live, whatever their race, gender, or economic status, whatever their immigration status or whatever language is spoken at home by their parents, and whatever their career goals." Of particular concern to the NSF was whether an uneven distribution of opportunities to learn science and mathematics might be contributing to unequal outcomes. It seems obvious that students won't learn what they are not taught, and that they won't learn well if they are not taught well. However, no comprehensive studies have investigated what various groups of students experience in their schools and classrooms; and no analyses have been performed that suggest how these experiences might restrict learning opportunities. Without such analyses, educators and policymakers have found it difficult to frame initiatives that might help achieve the NSF's goal.

The NSF therefore asked RAND to undertake a study of the way the nation's educational system distributes opportunities to learn mathematics and science among various groups of students. The inequalities documented here should be of interest to policymakers and educators who are concerned with improving both the processes and outcomes of mathematics and science education.

Some education observers resist considering children's learning opportunities in the absence of other, often implicit variables. For example, some who see schools as meritocratic institutions consider achievement itself as the principal mediator of opportunity, arguing that children who achieve more are better able to benefit from and more deserving of the limited resources that are available. Others explain opportunity, achievement, and participation as a function of mental capacity; for them, the most important opportunities are conferred at birth or before (i.e., they believe that some groups of children, because of racial or class-linked heredity, simply do not have the mental capacity to be very high achievers). While the attribution of lower achievement and participation to an entire group's supposedly lesser capabilities has been thoroughly discredited—and is clearly out of fashion—proponents of this viewpoint remain active, though their arguments may be more subtle than in the past.

Other observers believe that children's physiological history—particularly mothers' and children's nutrition and drug use—must be included in any discussion of children's opportunities to learn. Still others look to theories of cultural deprivation or to the nation's history of racial and/or class biases. Finally, some see inequalities as a regrettable but inevitable consequence of a shortage of high-quality educational resources and an attempt to use those resources in ways that will bring what they consider the highest educational return.

This study in no way attempts to discredit, endorse, or debate these viewpoints; they are merely acknowledged as having the capacity to shape the reading and interpretation of the findings reported here. Certainly they constitute an important context for understanding school practices. For example, the use of tracking and ability-grouping in mathematics and science stems from the widespread belief that children's intellectual differences are so great that students with different perceived ability levels need to be taught in separate classes and that much of the curriculum, especially at the secondary level, is not appropriate for many students. Many see the coincidence of these differences with students' racial and socioeconomic status as distressing, but not a matter over which schools have much control. Furthermore, many ignore the overall ineffectiveness of such grouping practices in increasing achievement.

Categorical differences in schooling opportunities are important, for both educational and political reasons. First, unequal learning opportunities provide some specific clues to how educational practices may help create and perpetuate differences in achievement and participation. Thus, the patterns that emerge suggest important targets for policies aimed at increasing students' educational outcomes.

Second, whether or not opportunities push a particular group of children toward higher achievement may not be as important a consideration as the fact our nation views equal opportunity as a democratic birthright. Yet the quality of the learning opportunities available to different categories of children relates strongly to the social and economic circumstances of children's families and communities. That such inequalities have no place in a democratic society is unarguable and should not be controversial.

SUMMARY

Widely published statistics document patterns of disproportionately low achievement and participation in science and mathematics by women, minorities, and the poor. These patterns are generating increasing concern as the nation's economic base shifts toward technology and the traditional pool from which scientific workers have been drawn (i.e., young white males) continues to shrink. Without substantial increases in the educational achievement and participation of currently underrepresented groups, the nation may not be able to meet its future scientific and technological needs. These human-capital issues converge with the long-standing policy objective of a fair distribution of economic and social opportunities. The specific policy issue of concern here is whether American schools give underrepresented and low-achieving groups of students an equal opportunity to participate and achieve in these increasingly important fields.

STUDY APPROACH

This report examines the distribution of science and mathematics learning opportunities in the nation's elementary and secondary schools. It addresses four key questions:

1. What science and mathematics are being taught to which students?
2. How are these subjects being taught?
3. By whom are they being taught?
4. Under what conditions are they being taught?

The educational system in the United States does not allocate opportunities directly to individuals; rather, it provides them to groups of students, first through schools and then through classrooms. We have examined opportunities that are available at different schools, opportunities available in different classrooms within schools, and finally, the participation of various groups of students in those classes and schools. We have considered not only differential opportunities associated with students' race, social class, and neighborhood, but also the uniquely school-bound distinction of ability-group, or "track," level. In brief, we have investigated whether different types of students have different opportunities to learn science and mathematics, and whether schools act on their judgments about students' academic abilities in ways that limit science and mathematics opportunities generally, and the opportunities of poor and minority students in particular.

Cross-sectional data about science and mathematics programs, teachers, and classroom practices in elementary and secondary schools obtained through the National Science Foundation's 1985-1986 National Survey of Science and Mathematics Education (NSSME) provided an unprecedented opportunity to describe the access of various groups to critical schooling elements. We have analyzed the distribution of various features of science and mathematics programs through cross-tabulations, correlational analyses, and analysis of variance. We have contrasted schools serving students of different racial, ethnic, and socioeconomic backgrounds, and classrooms enrolling various types of students. We have used multivariate analyses to isolate the effects of particular school and classroom characteristics, and separate classroom analyses within schools of various types. These analyses provide important information about whether and how the distribution of specific features of schools and classrooms may affect the learning opportunities of different students.

FINDINGS

During the elementary grades, the science and mathematics experiences of children from low-income families, African-American and Hispanic children, children who attend school in central cities, and children who have been clustered in "low-ability" classes differ in small but important ways from those of their more advantaged and white peers. By the time the students reach secondary school, their science and mathematics experiences are strikingly different.

The Distribution of Judgments About Ability

Assessments of academic ability, placement in different tracks or ability-grouped classes, and the reduced educational opportunities that characterize low-track classes often parallel race and social class differences. At schools with large concentrations of low-income and non-Asian minority students, disproportionate percentages of teachers judge their science and mathematics students to have low ability. At schools with racially mixed student bodies, the proportion of classes judged to be high-ability diminishes as minority enrollment increases, and minority students are more likely than their white peers to be placed in low-track classes. Thus, to the extent that placement in classes at different ability levels affects students' opportunities to learn—and the evidence from our study suggests that the effects are quite profound—minority students disproportionately suffer whatever disadvantages accrue to students in low-track classes.

The inequitable practices related to ability-grouping that we have identified in this study are commonly viewed as natural responses to differences in student aptitudes and achievements. But even if supposedly objective ability groupings appear logical, they are easily confounded with race and social class. Moreover, the differences in opportunities they provide actually *limit* instruction, rather than fine-tune it. Disparities in secondary school opportunites may reflect earlier conditions that have reduced the skills of disadvantaged students. However, we also see significant effects of race, social class, and locale on opportunities at the elementary level, where the cumulative effects of discrimination are less strong and where tracks are less predicated on prior achievement.

Access to Science and Mathematics Programs

With the exception of slightly greater amounts of time allocated to mathematics instruction in elementary schools with high concentrations of low-income and minority children, students from groups that as adults consistently achieve and participate less in science and mathematics have less access to science and mathematics curriculum. Low-income African-American and Hispanic students enrolled in secondary schools where they are the majority have less-extensive and less-demanding science and mathematics programs available to them. They also have fewer opportunities to take the critical gatekeeping courses that prepare them for science and mathematics study after high school—algebra and geometry in junior high school and calculus in senior high school. High-ability students at low-socioeconomic-status (SES), high-minority schools may actually have fewer opportunities than low-ability students who attend more advantaged schools. Moreover, overall differences in schools' science and mathematics programs are often compounded by inequalities in the opportunities available to various groups of students within schools. Students in low-track classes (disproportionately high percentages of whom are low-income and minority students) are far less likely than other students to be taking courses that emphasize traditional academic science and mathematics content. Although the differences are, in part, symptomatic of earlier conditions that fail to prepare disadvantaged students for rigorous courses, the net effect is that economically disadvantaged and minority students have considerably less access to the knowledge considered necessary either to pursue careers in science and mathematics or to become scientifically literate, critical-thinking members of an increasingly technological workforce.

Access to Qualified Teachers

Several measures of teacher qualifications make clear that low-income and minority students have less contact with the best-qualified science and mathematics teachers. The frequency with which teaching vacancies occur and the difficulty principals have filling vacancies with qualified teachers vary considerably among different types of schools. Teacher shortages appear most detrimental to low-income and minority students.

Most elementary and secondary school principals are fairly satisfied with the caliber of their science and mathematics teaching staffs, but principals of racially mixed and high-minority schools more often complain that lack of teacher interest and/or inadequate preparation to teach causes serious problems at their schools. Principals at schools enrolling large concentrations of low-income or minority students or at schools in inner cities also report that fewer of their teachers are highly competent. Teachers are even less sanguine. Teachers at high-poverty, high-minority, and inner-city schools report most frequently that lack of teacher interest or insufficient background poses problems for science and mathematics instruction. Moreover, secondary teachers in inner-city and rural schools and schools enrolling large concentrations of low-income children are less confident about their own science or mathematics teaching than teachers in more advantaged schools.

Evidence about teachers' formal qualifications reveals many of the same patterns. In this study, we found scant evidence of differences in certification status, academic background, and teaching experience among elementary teachers working in different types of schools (possibly because the nature of quality differences is hard to quantify at the elementary level), but we found substantial differences at secondary schools of different types. Schools whose students are predominantly economically advantaged and white and suburban schools employ teachers who are, on average, more qualified. Students attending these schools have greater access to science and mathematics teachers who are certified to teach their subjects, who hold bachelor's or master's degrees in those subjects, or who meet the standards set by professional associations.

Similarly, we found few differences in the qualifications of those teaching science and mathematics classes at different track levels at the elementary level, and substantial differences at the secondary level. Junior and senior high school students in low-ability classes are being taught by teachers considerably less well qualified than those teaching other levels. Nearly all types of secondary schools tend to place their least qualified teachers with low-ability classes and their most qualified teachers with high-ability classes. However, not all low- and high-track classes are equal, because of differences in the teacher pools available. In schools with less-qualified pools, teachers of low-track classes are less well qualified than those in schools with generally more qualified staffs. Students at the least advantaged schools must compete (through their class assignments) for teachers who are certified to teach mathematics and science or who have bachelor's degrees in these fields. In schools where teachers are generally more qualified, the sorting of teachers is evident on more subtle or higher-level qualifications—teachers' perceptions of themselves as "master" teachers, years of teaching experience (which may represent either seniority or political clout in the school), and the holding of master's degrees. As a result, high-track students in the least advantaged schools are often taught by teachers who are less qualified than those teaching low-track students in more advantaged schools.

Access to Resources

Students' access to science and mathematics facilities and equipment appears to be similarly unequal. Students in low-income, high-minority schools have less access than students in other schools to computers and to staff who coordinate their use in instruction, to science laboratories, and to other common science-related facilities and equipment. Additionally, more principals and teachers at less-advantaged schools report that resource problems interfere with science and mathematics instruction. Finally, instruction in low-ability classes appears to be further constrained by science and mathematics texts that most teachers judge to be of lower quality.

Classroom Opportunities

The curricular goals that teachers emphasize and the instructional strategies they use also differ in ways that further confirm the unequal opportunities of disadvantaged, minority, and inner-city students. Teachers serving large proportions of these students place somewhat less emphasis on such essential curriculum goals as developing inquiry and problem-solving skills. These disadvantages are compounded by differences in the curricular emphases in classes at different track levels, with low-ability classes the object of less teacher emphasis on nearly the entire range of curricular goals. Similar double-layered differences appear in classroom instruction. Teachers in schools with large concentrations of low-income and minority students are less likely to promote active involvement in mathematics and science learning. Students who are classified as average- and low-ability are disadvantaged in their access to engaging classroom experiences and teacher expectations for their out-of-school learning. Consequently, unequal access to science and mathematics curriculum goals is exacerbated by discrepancies in instructional conditions in classrooms.

These findings do not suggest that schools are differentiating science and mathematics curricular goals and instructional strategies in ways that are appropriate to the needs of students at different ability levels. On the contrary, students in low-track classes simply have less exposure to the teaching goals and strategies that are most likely to generate interest and promote learning among students at all achievement levels. Since low-income and minority students are disproportionately assigned to low-track classes, these differences further disadvantage these groups.

IMPLICATIONS

Our evidence lends considerable support to the argument that low-income, minority, and inner-city students have fewer opportunities to learn science and mathematics. They have considerably less access to science and mathematics knowledge at school, fewer material resources, less-engaging learning activities in their classrooms, and less-qualified teachers. These inequalities are linked to both characteristics of the schools and characteristics of the classrooms. Because schools judge so many low-income and minority students to have low ability, many of these students suffer from being in classrooms that offer less, even if their schools, as a whole, do not. Moreover, our findings are likely to be equally relevant for subject areas other than mathematics and science. The differences we have observed are likely to reflect more general patterns of educational inequality. As such, the implications of these findings extend beyond science and mathematics.

Our findings raise complex educational and ethical issues. Even though the data from this study do not link unequal opportunities directly to differences in achievement and participation, they provide some important and specific clues about how educational practices may help create and perpetuate these differences. But whether or not equal opportunities push a particular group of children toward higher achievement, our nation rejects the view that we should provide less to those who are less advantaged or less able. Yet inner-city schools serving large concentrations of children from poor families or African-American and Hispanic minorities often lack the political clout to command resources equal to those of other schools. Teachers often view these schools as less-desirable places in which to teach, partly because of the economic and social disadvantages that shape their students' lives. Also, these schools often pay less than surrounding suburban schools and offer poorer working conditions.

Within schools, educators believe they base decisions about who teaches what science and mathematics, to whom, how, and under what conditions on egalitarian and educationally sound criteria. But the processes and outcomes of tracking are complex, subtle, often informal, and incremental. Although the decisions are usually well-intentioned, considerable evidence suggests that tracking, especially at secondary schools, fails to increase learning generally and has the unfortunate consequence of widening the achievement gaps between students judged to be more and less able. Although schools may *think* that they ration good teaching to those students who can most profit from it, *we find no empirical evidence to justify unequal access to valued science and mathematics curriculum, instruction, and teachers.*

Moreover, the inequalities are not likely to be either self-correcting or easily changed by policymakers or educators. As long as high-quality educational opportunities are scarce and strategies for teaching diverse groups of students are largely untested, powerful constituencies of advantaged communities and parents will seek to preserve the educational advantages they now have. Consequently, it will be necessary for policymakers and educators to seek strategies that will ameliorate present inequalities and at the same time improve the science and mathematics education provided to all students. A multiple-strategies approach seems most appropriate for this complex and controversial policy issue.

RECOMMENDED STRATEGIES

Call Attention to the Problem

Policymakers would do well to expand their efforts to fuel public concern about educational opportunities as well as outcomes. Making better and more evenly distributed learning opportunities a focus of national concern can help clarify means for addressing issues of educational quality, future economic competitiveness, and social and economic justice. Strong advocacy from Washington and the state capitals would go a long way toward establishing a receptive climate for policies and practices aimed at both improving opportunities and distributing them more fairly.

Generate Additional Resources

Policymakers must seek new resources through new public funding, creative uses of existing funding, and new alliances with the private sector. And these resources should be accompanied by policies that change priorities for their allocation. New resources for materials and staff should go first to those schools with the greatest need—those that lag behind in computers, laboratories and materials, and well-qualified teachers. Like other affirmative-action strategies, however, policies aimed at providing new resources for these schools will confront political opposition to what may be seen as preferential treatment. The determination to ward off that opposition is often more easily sustained at the federal level. Nonetheless, state and local policymakers must also frame such farsighted policies.

Distribute Resources and Opportunity More Equitably

Many states are currently renewing their efforts to equalize funding levels across districts and schools. Such efforts, if successful, could provide the resources low-income schools need to purchase the facilities, materials, and staffing they now lack. But financial incentives may need to be altered to prevent good teachers from abandoning schools that serve low-income and minority students.

Policies are also needed that encourage a more equitable distribution of resources and opportunities *within* schools. For example, the federal government, states, local education agencies, and universities can all initiate programs aimed at developing new knowledge and building staff capacity to work effectively with diverse groups of students.

New school organizational schemes must be developed. These might include flexible staffing patterns such as teams of teachers sharing responsibility for diverse groups of students and/or staggered working hours to provide some teaching staff extra instructional time after school or in the evening for students requiring additional help. Other arrangements could involve more flexible use of resources from categorical programs. But if schools hope to make greater science and mathematics learning opportunities accessible to diverse groups of students, they will also need to redesign science and mathematics curriculum and instruction. Such curricular developments will help ensure that any move away from ability-grouped classes will be accompanied by higher-quality instruction for all students. Perhaps most important, improved curriculum and instruction should bolster the skills of currently disadvantaged children early on, so that they can more easily claim access to rigorous mathematics and science courses in junior and senior high school.

Hold States, Districts, and Schools Accountable for Equalizing Opportunity

Finally, given the difficulty and the potential political disincentives to equalizing educational opportunities, federal, state, and local efforts to reach this goal should be carefully monitored. As long as states view public accountability schemes as tools for encouraging local efforts to increase student outcomes, equalizing opportunities should be a part of what districts and schools are held accountable for. Educational data systems should be designed to report indicators of school resources, curriculum, teachers, instructional conditions, and outcomes by student race and SES. Such indicators could provide insights into how new educational policies could interrupt the patterns of unequal opportunities. Moreover, the public accounting could inform and energize communities and parents who may not otherwise realize that their children are getting less. Such monitoring efforts should be supported by a hierarchy of financial incentives to develop programs for equalizing opportunity, beginning at the federal level and extending to states, communities, and schools.

ACKNOWLEDGMENTS

Although responsibility for the analyses and interpretations in this study remains with the authors, the report has been enhanced by the generous involvement of a number of fine colleagues. Shirley Malcom and Audrey Champagne of the American Association for the Advancement of Science provided helpful comments on the initial questions and design of the study, as did Leigh Burstein of the University of California, Los Angeles, Thomas Romberg of the University of Wisconsin, and Kenneth Sirotnik of the University of Washington. Iris Weiss of Horizon Research provided ongoing guidance regarding the design and use of the 1985-1986 National Survey of Science and Mathematics Education (NSSME). Linda Darling-Hammond of Columbia University and RAND colleague Arthur Wise provided insightful reviews. Richard Berry and Ronald Anderson of the National Science Foundation gave encouragement and support. Finally, Janet DeLand lent a fine editor's hand to the final report.

CONTENTS

FIGURES

TABLES

I. THE DISTRIBUTION OF OPPORTUNITY

In 1983, the National Science Foundation (NSF) set an ambitious goal for precollege science and mathematics education: to provide "high standards of excellence for all students—wherever they live, whatever their race, gender, or economic status, whatever their immigration status or whatever language is spoken at home by their parents, and whatever their career goals" (National Science Board (NSB), 1983:12). But the disproportionately low achievement and participation in science and mathematics of women, minorities, the poor, and high school students who are not in college-preparatory programs reveals clearly that this goal is not being met.[1]

The lack of achievement and participation by these groups has generated considerable concern as the nation's economic base shifts increasingly toward technology. This concern is heightened by demographic projections showing that the traditional pool from which scientific workers have been drawn, i.e., young white males, is shrinking. Future cohorts of workers will comprise increasing proportions of non-Asian minorities—groups that traditionally have not entered scientific and technological fields. These changes raise a number of specific policy questions: How can we ensure an adequate future supply of highly trained mathematicians, scientists, and engineers? How can we provide the general labor force with the knowledge and skills needed for technological work? How can we attain the level of scientific literacy necessary for responsible, democratic decisionmaking about scientific and technological matters? There are no clear-cut answers to these questions. However, many observers suggest that if the educational achievement and participation of minorities do not increase substantially, the nation will not be able to meet its scientific and technological needs.

These human-capital issues have converged with the long-standing policy objective of fair distribution of economic and social opportunities. As technology becomes increasingly central to work and national life, the achievement of women and minorities in science and math-

[1]These discrepancies have been detailed in several reports over the past five years, including American Association for the Advancement of Science (AAAS), 1984; Achievement Council, 1985; Berryman, 1983; Chipman and Thomas, 1984; Darling-Hammond, 1985; National Alliance of Black School Educators (NABSE), 1984; National Science Board, 1987; National Science Foundation, 1988; Oakes, 1990; and Task Force on Women, Minorities, and the Handicapped in Science, 1988.

ematics will be a primary factor in the ability of these groups to compete for employment, wages, and leadership positions. While not all students have the interests or aptitude to become scientists or mathematicians, the disparities for African-American and Hispanic minorities and the poor are so great that considerable science and mathematics talent is undoubtedly being lost from these groups. Moreover, many minority and poor students are failing to reach even the levels of mathematics and science literacy believed to be necessary for knowledgeable participation in an increasingly technological society. Minorities have made important progress toward closing the achievement gap in the past two decades, but appalling disparities in school achievement and occupational status remain.

The NSF was particularly concerned with the possibility that an uneven distribution of opportunities to learn science and mathematics might be contributing to unequal outcomes. It is obvious that students will not learn what they are not taught and that they will not learn well if they are not taught well. However, no comprehensive descriptions of what various groups of students experience in their schools and classrooms have been available, nor have analyses been performed to suggest how these experiences might restrict learning opportunities. Without such analyses, educators and policymakers have found it difficult to frame initiatives that could help to achieve the NSF's goal.

This report responds to these concerns by examining the distribution of science and mathematics education in the nation's elementary and secondary schools. It provides information that should help to answer four key questions: What science and mathematics are being taught to which students? How? By whom? And under what conditions? The report has three broad objectives:

1. To document the differences in science and mathematics curriculum, resources, classroom activities, and teacher quality among various groups of students in the nation's schools.
2. To provide insights into how those differences might shape the learning opportunities of groups that typically have low levels of achievement and participation in science and mathematics.
3. To explore the implications of these findings for precollege science and mathematics education policy and practice.

Drawing primarily on data from the 1985-1986 National Survey of Science and Mathematics Education (NSSME), we explore whether access to science and mathematics curriculum, resources, instructional activities, and teachers relates to (1) characteristics of the school a student happens to attend, (2) characteristics of the classroom in which a student is enrolled, or (3) characteristics of school and classroom combined.

Our analyses reveal clear and consistent patterns of unequal opportunities to learn mathematics and science. During the elementary grades, the science and mathematics experiences of large numbers of low-income children, African-American and Hispanic children, children who attend school in central cities, and children who have been judged to have "low ability" differ in small, but important ways from those of their more advantaged or white peers. By the time these students reach secondary school, the differences are striking. Low-income, minority, and low-ability students have considerably less access to science and mathematics knowledge; they have fewer material resources available to help them learn these subjects; their classrooms offer less-engaging learning activities; and their teachers are less-qualified. These differences can be traced to characteristics of both the schools in which different groups of students are clustered and the classrooms in which they are taught. Because school officials judge so many low-income and minority students to have low ability, many of these students suffer the double disadvantage of being in schools that have fewer resources and classrooms that offer less access to knowledge.

ORGANIZATION OF THE REPORT

The remainder of this section describes a conceptual framework for understanding the distribution of opportunities. This framework suggests that distributional analyses must consider a comprehensive set of school, classroom, and student characteristics, and that the best way to assess individuals' opportunities is by examining the schools and classrooms in which particular groups of students are clustered. Finally, it describes the data and methods used in the study and their limitations. Section II describes how students' race and social class characteristics overlap with schools' assessments of their abilities and their placement in various types of science and mathematics classes. Section III examines the distribution of science and mathematics curricula, as evidenced by the types of courses schools offer. Section IV examines teachers' experience and qualifications and assesses how the distribution of these factors may influence students' learning and participation. Section V analyzes the allocation of science and mathematics resources to schools of various types. Section VI considers classroom processes—the emphasis teachers give to various curricular objectives, the instructional activities they include in lessons, and how they use classroom time. Finally, Section VII discusses the implications of the distributional patterns found in the study. Our focus on the *specific* dimensions of science and mathematics opportunities—resources, teachers, curriculum, and instructional practices—enables us to evaluate policies and practices that are likely to remedy discrepancies.

DIMENSIONS OF THE DISTRIBUTION OF OPPORTUNITY

Our educational system does not allocate opportunities directly to individuals; rather, it allocates them to groups of students—first through states and school districts, and then through schools and classrooms. Consequently, distributional studies require comprehensive analyses at both the school and classroom levels, and the findings will undoubtedly reflect the resources made available by states and districts. It is necessary to describe, first, differences in the resources, instructional conditions, and teachers available to schools and, second, how schools distribute those resources to different classrooms. While this departs somewhat from the usual approach of focusing on individuals or groups to explain how their opportunities differ, this institution-based rather than individual- or group-based approach has several advantages. First, the clustering of students in schools and classrooms strongly influences the opportunities that they enjoy. Students' access to knowledge, resources, teachers, and classroom processes is shaped by the characteristics of the schools and classes in which they are enrolled. Moreover, this approach enables us to examine the availability of opportunities at several grade levels. This adds important information, since what students actually experience in their science and mathematics classrooms, from the earliest grades through senior high school, will cumulatively influence both what they learn and whether they continue to participate in the pre-college mathematics and science pipeline.

At the school level, data are needed on the programs offered and the human and material resources available to deliver them. The courses that make up secondary schools' science and mathematics curricula suggest the programs' breadth, depth, and extent. In elementary schools, the amount of instructional time spent in science and mathematics is critical. In both elementary and secondary schools, the ability composition of classes (heterogeneous or homogeneous low-, average-, or high-ability) signals whether students have differentiated learning opportunities.[2,3] Grouping of students within classrooms may indicate differential treatment, particularly in the elementary grades. Science and mathematics opportunities at a school are also shaped by the quality of the teaching staff, materials and equipment available, and obstacles teachers face as they teach these subjects.

School characteristics determine the conditions under which classroom teaching and learning occur, but how well students learn and how long they sustain an interest in the subjects they are taught are most influenced by day-to-day classroom experiences.[4] In classrooms, we need to examine teacher quality, curriculum, and instructional activities and how these factors affect students' access to learning opportunities. Similar course titles at secondary schools can represent quite different learning experiences, just as similar amounts of time allocated to instruction in elementary school subjects can mask substantial differences in how that time is spent. What students experience in classrooms is determined largely by teachers' instructional goals and objectives; the knowledge and processes teachers make available; the books, materials, and equipment teachers use; the classroom learning activities teachers arrange; the quality of teachers' background, training, and experience; and the support and resources available to teachers.

Distributional issues are not confined to *general patterns* of variation across the entire population, however; policymakers are increasingly concerned with the distribution of opportunities to *particular groups*, defined by where they live, race, gender, or economic status; immigration status; the language spoken at home; or their career goals (NSB, 1983:12). Therefore, it is necessary to analyze the relationship between student composition of schools and classrooms, the types of communities in which the schools are located, and the resources and instructional conditions they provide.

---

[2]Throughout this report, we use the terms *ability grouping* and *tracking* interchangeably to mean the clustering of students who are judged to be similar in their academic ability into classes for instruction.

[3]At the elementary level, however, further differentiation often occurs when teachers form ability-based instructional subgroups.

[4]See Barr and Dreeben (1983) for a discussion of the importance of a multilevel approach for understanding how schools work to produce student learning.

In 1983, the NSB also suggested that distributional inequalities are likely to be linked to beliefs about the abilities of students to learn and asserted that "the opportunity to learn mathematics, science, and technology is at present not fairly and evenly provided to all students, and that in the past, such inequalities have resulted from the failure to recognize and develop potential talent, from inadequate educational programs in some communities or for certain groups of students, and from the erroneous belief that many students lacked the ability to learn mathematics and science" (NSB, 1983:13). Thus, distributional analyses must consider the effects not only of race and social class but also of teachers' judgments about students' abilities.

Ability classification is most useful in distributional analyses as a descriptor of the *class as a whole*, since it is a potentially important mediator of the resources and opportunities provided to *all* the students enrolled in the class. Teachers (particularly at the secondary level) typically make curricular and instructional decisions at the *class* rather than the *individual* level. Even in elementary schools, many teachers do not form instructional subgroups in mathematics, and teachers rarely use within-class ability groups for science instruction (Oakes, 1985; Slavin, 1987). Therefore, in nearly all cases, the teacher's perception of a class's ability plays an important role in deciding what and how to teach it.

While most people (including many educators) assume that students will learn better if they are grouped together with those who have similar capabilities, research has shown that putting children into separate classes to accommodate their differences from their earliest school years is neither necessary nor very effective. Tracking does not work well for students in the low- and middle-ability groups, who experience clear and consistent learning disadvantages. Perhaps more surprising, tracking does not necessarily promote achievement for high-ability children either: Many studies show that highly capable students do as well in mixed-ability classes (Gamoran and Berends, 1987; Oakes, Gamoran, and Page, in press; Slavin, 1987; Slavin, 1990).

It is also well-established that tracking separates students by race and social class. African-American and Hispanic students are disproportionately assigned to low-ability classes and to non-college-preparatory high school programs, as are students from low-income families.

However, tracking is not only ineffective and segregative; it also leads to very unequal learning opportunities (see, e.g., Oakes, 1985, 1987). Students in different groups and tracks have *access to very different types of knowledge*—those in high-track classes are more likely to study rich and meaningful topics and skills, while those in low-track classes get a low-level curriculum dominated by exercises, workbooks, and commercially produced basic-skills kits. Small-scale studies show these differences exist at nearly every grade level and in nearly every subject.

There are also important *differences in classroom instruction.* Students in higher-level classes spend more time on learning activities and less time on discipline, socializing, or class routines. Teachers of these classes usually teach more enthusiastically, and they make their instruction clearer. They tend to organize tasks better and give children a greater variety of learning activities, and they expect their students to spend more time doing homework. In contrast, students in low-track classes more often feel excluded from class activities and find their classmates unfriendly. Problems and arguments interrupt classes more frequently. Moreover, students in low-ability classes seem apathetic; being more likely to fail, they may feel that they risk much more by trying hard and giving the appearance that they care.

What prior work shows, then, is that tracking produces fundamental schooling inequities. Students who need more time to learn get less; those who have the most difficulty learning experience less good teaching. In contrast to what is commonly assumed—that students are assigned to various ability-grouped classes because they belong there and that those classes serve their educational needs—prior research suggests a very different conclusion. Designations of "ability" are suspect (given their links with race and social class), even though they may relate to students' prior school performance; and "ability-based" inferences about students' curricular and instructional needs are often wrong.

These findings and conclusions are particularly important for understanding the underparticipation of minorities and low-income students in science and mathematics. Since patterns of enrollment and placement in ability-grouped classes have been found to be linked to race and social class, and since combinations of race, class, and track placements relate to the learning opportunities provided in classrooms, four questions are particularly important:

1. What are the differences in the mathematics and science opportunities schools provide based on judgments about students' ability?

2. Are these differences likely to constructively accommodate individual differences in mathematics or science abilities?
3. How do judgments about ability and placements in various classes relate to other student background characteristics (e.g., race, class, and gender)?
4. What are the combined effects of race, class, and tracking on students' opportunities to learn mathematics and science?

To answer the fourth question, we must compare the resources and opportunities various types of schools (defined by socioeconomic status (SES), racial composition, locale, etc.) provide to students the schools perceive as most able, average, and least able.

In sum, distributional analyses must assess several kinds of distributional differences: those that seem to be a consequence of the schools students happen to attend (e.g., do students attending inner-city schools have different coursetaking options than those in suburban schools?), those that come about because of the classes in which students are enrolled (e.g., are students in high-track classes taught by different types of teachers than those in low-track classes?), and those that follow from combined school and classroom factors (e.g., do students in high-ability classes in affluent schools experience instruction similar to that received by their high-track peers in schools with high proportions of low-income students?).

This framework for understanding the distribution of opportunities to learn science and mathematics looks at opportunities that are available at different schools, opportunities available in different classrooms within schools, and finally, the participation of groups of students in those classes and schools. To the more established concerns for differential opportunities associated with students' race, social class, and neighborhood, we add the uniquely school-bound status distinction of perceived ability level. In brief, we ask whether students have different opportunities to learn science and mathematics, and if so, whether these opportunities are associated with students' race, class, and neighborhood. We also ask whether schools act on their judgments about students' abilities in ways that limit science and mathematics opportunities generally, and the opportunities of poor and minority students in particular.

STUDY APPROACH

We approach our inquiry of the distribution of opportunities to learn science and mathematics from two perspectives, one reflecting research on school and classroom factors that relate to the participation and achievement of women and minorities in science and mathematics,[5] and one reflecting our analyses of data from the NSSME, which provide an unprecedented amount of information on the access of various groups to a whole constellation of critical schooling elements. The juxtaposition of these two perspectives enables us to suggest whether and how the distribution of specific features of schools and classrooms observed in the data are likely to affect the learning opportunities of various groups.

The Database

The NSSME was administered to principals and teachers in a national probability sample of 1,200 public and private elementary and secondary schools. The sample was designed to allow estimates of several dimensions of schools, teachers, and classroom practices in science and mathematics nationwide, as well as estimates for various subpopulations defined by region, type of community, and school type. Approximately 6,000 teachers of science and mathematics were randomly selected from within the sampled schools. One of the mathematics or science classes of each secondary teacher was randomly selected as the focus for the class-specific items.

Separate instruments were fielded for secondary and elementary school principals, secondary science and mathematics teachers, and two groups of elementary teachers. One elementary teacher sample responded about science teaching; the other responded about mathematics. The questionnaires focused on the school contexts of mathematics and science education, descriptions of programs in these subjects, and specifics about curriculum and classroom instruction. Data were also collected on the science and mathematics background, training, experience, and attitudes of teachers and administrators. Administrators reported the race and socioeconomic backgrounds of the students attending the schools; teachers detailed the race, gender (and gender breakdowns within racial and ethnic groups), and ability levels of students in their classes. The data thus permit analyses of between- and within-school differences for various groups of students

[5]This literature is reviewed in Oakes, 1990.

(defined by race, social class,[6] and track levels, both separately and in various combinations) at three levels of schooling (elementary, junior high/middle school, and senior high).[7]

Analytic Approach

We use cross-tabulations, correlational analyses, and analysis of variance to examine the distribution of mathematics and science programs, teachers, facilities and equipment, and classroom experiences. We contrast schools serving students of different racial and ethnic groups and with varying socioeconomic backgrounds, and classrooms enrolling various types of students. We use multivariate analyses to help sort out the effects of school and classroom characteristics, and in some cases, we conduct separate classroom analyses within schools of various types.[8]

LIMITATIONS OF THE STUDY

While our findings provide new insights into the distribution of science and mathematics opportunities and the implications of that distribution for the participation and achievement of currently underrepresented groups, the study also leaves many questions unanswered. First, although research increasingly points to the importance of school and classroom processes for learning, our understanding of how various resources and opportunities enhance or constrain students' success is far from complete. Moreover, research on the relevance of particular school and classroom features to the success of women, minorities, the poor, and those identified as low-ability is far from unequivocal. Therefore, our selection of dependent variables for analysis was based on a less-than-perfect understanding of which features of schools and classrooms are most important to examine.

---

[6]The NSSME data provide SES information only at the level of the school. Therefore, we analyze classrooms of various racial/ethnic and ability compositions as "nested" within schools having different SES levels. Using this approach, we can examine how curriculum, resources, and classroom activities in low-track science and mathematics classes in low-SES schools or schools with all or predominantly nonwhite enrollments compare with similar classes in schools with different student population characteristics.

[7]For a more detailed discussion of of the survey sample designs, survey populations, and weighting process, see Weiss (1987).

[8]Except where noted, the General Linear Models analysis-of-variance program was used for the analyses of school- and class-level variables.

Moreover, the NSSME data contain no information about students' achievement. Consequently, we can draw no conclusions about how the distribution of learning opportunities relates to students' performance in science and mathematics. However, such relationships have been found in analyses of other datasets.

Second, the NSSME data cannot answer questions about certain important features of mathematics and science education. For example, girls' motivation to study mathematics appears to be adversely affected by several classroom characteristics, e.g., lack of opportunity for decisionmaking, emphasis on whole-group teaching and drill, less individualized instruction, concern with discipline and control, use of explicit and public criticism, and stress on competition (Eccles, MacIver, and Lange, 1986; Fennema and Peterson, 1986; Lockheed, 1984; Lockheed, 1985; Parsons, Kaczala, and Meece, 1982; Peterson and Fennema, 1985). African-American and Hispanic students, too, may be disadvantaged under such classroom conditions (Armstrong, 1980; Kagan, 1980; Slavin, 1983). Both women and minorities have been shown to be more likely to persist in mathematics and science if they see these subjects as interesting, connected to everyday life, and relevant to their future careers (Casserly, 1979; Chipman and Thomas, 1984; Creswell, 1980; Fennema and Sherman, 1977; Lantz and Smith, 1981; Maccoby and Jacklin, 1974; Tobin and Fox, 1980).

Survey data simply do not allow us to examine the subtleties of classroom interactions and learning experiences. When possible, however, we use variables in the NSSME data set that suggest the presence or absence of classroom characteristics that appear to influence girls' and minorities' motivation and achievement in mathematics. These include the time spent on whole-class versus small-group or individualized instruction, the use of discussion as a teaching strategy, and teachers' emphasis on developing an interest in mathematics and science, becoming aware of the usefulness of mathematics and science, and seeing the career relevance of these subjects. The data do permit us to examine the distribution of these features among large numbers of classrooms enrolling various groups of students and the extent to which more encouraging conditions are present in classrooms of particular interest, e.g., those with large percentages of minorities.

Third, the NSSME survey data do not reflect the differences in the experience of various groups of students within the same classroom. Observational studies of teacher-student interactions and other, more subtle classroom features often find differential treatment within classrooms (for example, by student race, gender, or ability). This is not a serious problem for our study, however, since most of the classroom variables included in the NSSME data reflect either tangible resources (computers, calculators, funding, teacher qualifications, class size) or instructional decisions teachers typically make at the class level (such as how to deal with homework). Except possibly in upper elementary mathematics classes, where instructional subgroups may be formed according to ability levels, most of these factors will affect all of the students enrolled fairly evenly. Earlier work suggests that little individualization occurs within classrooms (e.g., Goodlad, 1984), so we can fairly safely assume that the broad patterns of experience reported in these data reveal class experiences shared by the students enrolled.

Fourth, we are unable to examine distributional differences related to gender. While we did not expect to find predominantly male or female mathematics and science classrooms at the elementary and junior high school levels, we thought we would find them at the senior high school level, since many classes at this level are optional. However, such distinct enrollment patterns did not appear in our analyses, so we were not able to explore whether there are systematic gender differences in access to resources and classroom experiences.

Other limitations that relate to particular constructs or variables are discussed in the context of specific analyses in subsequent sections.

II. THE EFFECTS OF STUDENT CHARACTERISTICS ON OPPORTUNITY

Relating students' race, social class, neighborhood, and career goals to resource allocation and instructional conditions is complicated, because while these classifications are individually important, they are also inseparable—no student has only one of them. In this section, therefore, we first consider the interrelatedness of these characteristics. We then use data from the NSSME to argue that the judgments schools make about students' ability should also be considered an equally inseparable characteristic that influences students' access to educational opportunities.

THE INSEPARABILITY OF STUDENT CHARACTERISTICS

Every student is a member of a racial group, a social class, and a type of community, and each can be identified as having a particular ability status (e.g., low, average, or high). In the United States, these various memberships and categories cluster.

African-American and Hispanic students[1] tend to be clustered in inner-city schools. Low-income[2] minority students are also concen-

[1]Our analyses of minority students focus on African-Americans and Hispanics. The small number of Native Americans in the sample does not permit us to extend our findings to that group, although we expect that their experiences parallel those of the African-Americans and Hispanics. We did not include the Asian students in our minority category, since our purpose was to describe the experiences of minority groups that are underrepresented in science and mathematics. We grouped Asian students with whites in our analyses; separate analyses excluding Asians did not yield different results, however, because of the small size of this group.

[2]At the school level, we created variables for SES based on principals' reports of school characteristics. Because recent work has established its importance for a number of schooling outcomes (Orland, 1988), the concentration of poor students was used as the basis for our SES measure, with schools reporting low concentrations further differentiated by the proportion of students from wealthy backgrounds. We are well aware that determining school and student SES is always difficult with surveys, and completely satisfactory proxies are rarely found. Our questionnaires asked principals to list the percentage of students with parents in the following occupational categories: (1) professional and managerial; (2) sales, clerical, technical, or skilled workers; (3) factory or other blue collar workers; (4) farm workers; (5) persons not regularly employed; (6) persons on welfare. We felt that while principals were likely to be fairly accurate in estimating percentages of students from families at the high and low ends of this occupational scale, they would be less able to discriminate among those

trated in other urban and rural schools, but they are not often in the majority there.[3] On the other hand, middle- and upper-income white students are concentrated in and form a majority in suburban schools.[4] Most rural schools tend to be populated by white students, most of whom are neither very poor nor very well off. Nevertheless, as Table 2.1 shows, there are exceptions. Noticeably absent from the exceptions are all-minority suburban schools and predominantly minority high-wealth schools. Moreover, disproportionately more African-American and Hispanic minority students, poor students, and inner-city-school students are classified by schools as being low in academic ability and not likely to attend college. (We shall look more carefully at these relationships later in this section.)

Not only do these patterns exist, the clustering of particular categories is not serendipitous. Underlying social, political, and economic conditions provide substantive explanations for the fact that so many African-American and Hispanic youngsters are low-income, are living in inner-city neighborhoods, and are perceived to be less able than students from other backgrounds. These same dynamics also help to explain why such a large proportion of whites are economically better off, live in more affluent urban and suburban communities, and are more likely to be considered academically able.

While it has been argued that researchers should treat students' characteristics as inseparable, interactive influences on educational and occupational attainments (Grant and Sleeter, 1986), little is known about the simultaneous or combined effects of these character-

---

parents who held mid-level jobs (categories 2 and 3). Moreover, we considered "farm workers" an ambiguous category, which principals might interpret to include a wide range of occupational levels, from poor, itinerant field hands to owner-operators. We therefore constructed our SES categories from responses to the extreme categories (1, 5, and 6), devising measures that reflect the extent to which wealth and poverty are represented. The following school SES categories were defined: (1) high poverty (more than 30 percent of the students have parents who are either unemployed or on welfare); (2) moderate poverty (between 10 and 30 percent of the students have parents who are either unemployed or on welfare); (3) low poverty/low to moderate wealth (less than 10 percent of the students have parents who are either unemployed or on welfare, and no more than 30 percent have parents in professional or managerial occupations); (4) low poverty/high wealth (less than 10 percent of the students have parents who are either unemployed or on welfare, and more than 30 percent of the students have parents in professional or managerial occupations).

[3]Categories of schools that differed in racial composition were defined as follows: (1) minority—less than 10 percent the of students are white; (2) mixed (mostly minority)—10 to 50 percent are white; (3) mixed (mostly white)—50 to 90 percent are white; and (4) white—more than 90 percent are white.

[4]We did not define new variables for determining a school's community type. Principals were asked to choose whether their school locations were best described as (a) inner city, (b) urban, but not inner city, (c) suburban, or (d) rural.

Table 2.1

SCHOOLS IN VARIOUS RACE, SES,[a] AND LOCALE CATEGORIES

(Percentage of all schools; N = 977)

	Inner City				Other Urban				Suburban				Rural				
Racial Composition	High Pov.	Mod. Pov.	Low Pov.	High Wlth.	High Pov.	Mod. Pov.	Low Pov.	High Wlth.	High Pov.	Mod. Pov.	Low Pov.	High Wlth.	High Pov.	Mod. Pov.	Low Pov.	High Wlth.	Total
0–10% white																	
Elementary	1.3	0.1	0.1	0	0.2	0.1	0.1	0	0	0	0	0	0.2	0.1	0	0	2.3
Secondary	1.3	0.5	0.1	0	0.4	0	0.3	0.1	0	0	0	0	0.4	0	0	0	3.2
10–50% white																	
Elementary	1.2	0.1	0.1	0	0.9	1.0	0.2	0	0.4	0.3	0.3	0	0.3	0.2	0.1	0	5.2
Secondary	0.4	0.2	0.1	0	0.8	1.8	0.8	0.1	0.3	0.5	0.2	0	0.2	0.7	0.3	0	6.6
50–90% white																	
Elementary	1.0	0.2	0.2	0	0.7	1.4	0.9	0.4	0.1	1.3	1.6	1.2	0.7	0.9	0.3	0.2	11.4
Secondary	0.6	0.9	0.6	0.1	0.5	2.3	2.2	1.2	0.5	1.9	2.2	2.5	0.8	3.2	1.5	0.3	21.3
90–100% white																	
Elementary	0.1	0	0.5	0.2	0.4	0.7	0.4	1.0	0.1	1.4	2.3	2.6	1.0	2.4	2.5	0.7	16.3
Secondary	0.2	0.1	0.5	0.1	0.3	1.2	1.9	1.4	0.3	2.2	4.3	7.4	0.8	5.8	6.1	1.1	33.9
Total	6.2	2.2	2.3	0.4	4.3	8.6	6.9	4.3	1.7	7.7	10.9	13.6	4.5	13.3	10.9	2.4	100

NOTE: Totals may not sum correctly due to rounding.

SOURCE: 1985-1986 NSSME.

[a]High poverty = more than 30 percent of students in school have parents who are either unemployed or on welfare; moderate poverty = 10 to 30 percent have parents who are unemployed or on welfare; low poverty and low-to-moderate wealth = less than 10 percent have parents who are unemployed or on welfare, and no more than 30 percent have parents in professional or managerial occupations; low poverty and high wealth = less than 10 percent have parents who are unemployed or on welfare, and more than 30 percent have parents in professional or managerial occupations.

istics. For example, while there are extensive data on the experiences of minorities *per se*, data for minority subgroups are rarely collected or analyzed separately. There is little data, for example, to document SES differences within minority groups, such as those between low-income and middle-class African-American students.

THE RELATIVE IMPORTANCE OF RACE AND SES

One of the most puzzling questions concerning student characteristics is whether race or SES has the greater effect on children's opportunities for achievement and participation. Because these characteristics are nearly always studied separately, there is scant evidence with which to address this question. Race has received the bulk of the attention, since racial discrimination has been more often the subject of court actions and federal programs. The experiences of poor children are often extrapolated from studies of racial differences, since many of the poorest children in the United States are African-American and Hispanic, and most of the affluent children are white. But inferring the status of poor children in this country from the circumstances of African-American and Hispanic children grossly and stereotypically oversimplifies matters. Race and class probably cannot be effectively disentangled in attempts to understand the lower rates of achievement and participation of African-American and Hispanic students, but, as Table 2.1 illustrates, the overlap is not perfect. A significant disadvantaged sector of the school population—children who are poor and white—is often overlooked, as are middle-class minority students.

The importance of social class status in itself has been demonstrated in recent analyses. High School and Beyond (HSB) data illustrate that with other school and student factors (including race and ethnicity) controlled, students' SES accounts substantially for differences in mathematics achievement (Rock, Braun, and Rosenbaum, 1985). Similar relationships are found in Scholastic Aptitude Test (SAT) scores (College Board, 1985).

These findings compel us not only to consider race and social class separately, but also to look at how combinations of student characteristics relate to schooling experiences. The objective should not be to determine which single characteristic is the *most* important; analyses that control for some characteristics in the attempt to determine the effects of any single characteristic may mask the fact that all of them are important. We need to know how race, social class, and neighborhood *independently* affect students' learning opportunities. But we must also remember the overlaps among these characteristics, and whenever possible, we must explore how *combinations* of characteristics affect students' chances to learn.

THE RELATIONSHIPS BETWEEN ABILITY JUDGMENTS AND OPPORTUNITY

An often neglected key to understanding the distribution of critical features of science and mathematics education is the close connection between educators' judgments about students' intellectual ability and the educational experiences that follow from those judgments. The NSB has suggested that inequalities may stem from the "failure to recognize and develop talent" and "the erroneous belief that many students [lack] the ability to learn mathematics and science" (NSB, 1983:13).

Growing bodies of evidence show that assessments of students' intellectual abilities play a major role in the differential allocation of school experiences (Gamoran, 1987; Guthrie and Leventhal, 1985; Lee, 1986; Oakes, 1983, 1985, 1987). Judgments about academic ability often lead to the segregation of students into separate elementary and middle-school classes and to enrollment in different senior high school courses. These placements, in turn, can mediate students' opportunities to learn. In elementary and middle schools, students who appear to be slow are often placed in lower-level groups or remedial programs; students who seem to learn more easily are placed in high-ability groups. Research evidence suggests that these placements influence the pace and content of instruction and contribute to achievement disparities (Barr and Dreeben, 1983; Hallinan and Sorensen, 1983). At the senior high school level, judgments about students' ability influence decisions about curriculum track enrollment—whether students take college-preparatory, general, or vocational courses of study. Track enrollment, in turn, is critical in coursetaking and achievement (Lee, 1986; Lee and Bryk, 1988; Rock et al., 1984, 1985), as well as in curriculum content, instructional practices, and classroom learning environments (Oakes, 1985).

Class and track placements and subsequent differences in learning experiences have traditionally been explained as appropriate, given the apparent differences in students' performance at school. However, some evidence suggests that the ways elementary schools define ability and respond to students may help to solidify students' perceptions of their prospects for achievement and may eventually exaggerate initial performance differences among them (Rosenholtz and Simpson, 1984). Slavin (1987) shows that the popular practice of whole-class ability grouping at elementary schools is ineffective in increasing student learning. Similarly, secondary school tracking does not increase schools' overall achievement. Some studies have found that tracking may raise achievement for high-track students slightly, but when these gains are found, students in low tracks suffer a corresponding loss. These findings raise the possibility that in trying to accommodate students' differences with different educational experiences, schools may actually limit most students' opportunities to learn.[5]

Judgments about students' ability and the corresponding placement in homogeneous ability groups can affect the quality of students' educational experiences and achievement; therefore, tracking must be considered an important mechanism for distributing science and mathematics opportunities.

RACE, SOCIAL CLASS, AND ABILITY CLASSIFICATIONS

Assessments of ability and placements in different classes appear to be particularly relevant to the educational experiences of poor and minority students, since the assessments often parallel race and class differences (Persell, 1977; Oakes, 1985; 1987; Rosenbaum, 1980). Low-income and minority elementary and middle-school students are more likely to experience initial learning difficulties; as a result, they are more likely to be judged as "low-ability" and placed in low-track and remedial classes or in special education programs (Persell, 1977; Rosenbaum, 1980; Slavin, 1987). Whites and upper-SES elementary students are more likely to be identified as able learners (and more likely to be considered "gifted and talented") and placed in enriched or accelerated programs (Darling-Hammond, 1985). In senior high school, African-Americans, Hispanics, and low-income students are enrolled more frequently in vocational and general programs, while whites and high-SES students are more frequently enrolled in academic programs (Rock et al., 1984, 1985).

There is a striking national pattern in the links between the tracking phenomenon and students' race and SES in the NSSME data, consistent with earlier research. Teachers at schools enrolling poor

[5]See Oakes, Gamoran, and Page (in press) for a review of this literature.

students and non-Asian minorities disproportionately judge their science and mathematics students to be of low ability.[6] In mixed schools, science and mathematics classes in which disproportionate numbers of minorities[7] are enrolled are highly likely to be perceived as low-ability classes.

Figure 2.1 shows that at both the elementary and secondary levels, schools in all four SES categories form classes that are homogeneous in ability, and that they do so to nearly the same extent.[8]

However, as Figs. 2.2 and 2.3 illustrate, despite the similarity in the percentages of homogeneous classes at both elementary and secondary schools, the *types of homogeneous groupings* differ quite dramatically in schools serving students of different SES levels.[9] Schools that have large concentrations of low-income students have a significantly greater percentage of low-ability groups. The percentage of low-ability classes drops and the percentage of high-ability groups increases significantly with higher school SES.

The minority enrollment at schools relates similarly to the patterns of ability groupings (Figs. 2.4, 2.5, 2.6). Figure 2.4 shows only small

---

[6]The ability or track level of a class represents the teachers' categorization of the class as including (1) a wide range of ability levels, (2) predominantly low-ability students, (3) predominantly average-ability students, or (4) predominantly high-ability students. We must emphasize two limitations of analyses of ability groupings in the NSSME data. First, we are dealing with perceived ability as a *characteristic of the class*—we can legitimately talk about opportunities available to students in low-track classes, etc., but we can say nothing about individual ability. Second, we are dealing with *teachers' perceptions* of ability—we cannot assume that ability means the same thing from teacher to teacher, or from school to school. Finally, while it is impossible to determine from the NSSME questionnaire whether a formal tracking system is in place at a school, the percentages of homogeneous groups reported by teachers agree very closely with the percentages of elementary schools reporting that they use whole-class ability grouping (Slavin et al., 1989).

[7]We used a variable created from principals' descriptions of school characteristics and teachers' descriptions of sampled classes to show the overrepresentation or underrepresentation of African-American and Hispanic students. We developed three categories for this variable: (1) disproportionately white—smaller percentage of minority students in the classroom than in the school (differences can range from 10 percent to a theoretical 100 percent); (2) proportionate—same percentage as in the school (within 10 percent in either direction); and (3) disproportionately minority—larger percentage of minority students than in the school (differences can range from 10 percent to 100 percent).

[8]Tthe most-affluent schools seem to group students homogeneously more frequently than the least-affluent schools, although the 5 to 10 percent difference is not statistically significant.

[9]Percentages in Figs. 2.2 through 2.6 do not sum to 100 because mixed-ability classes are excluded from these charts.

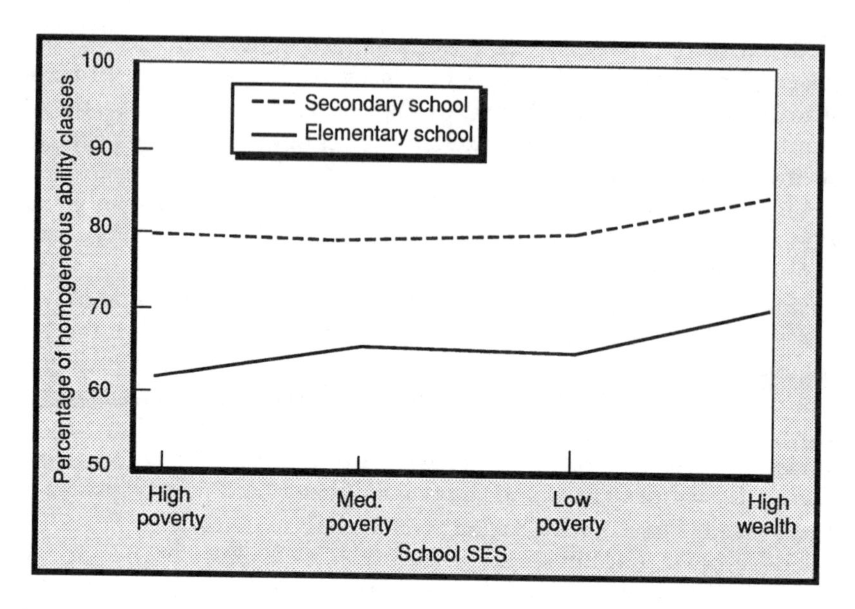

Fig. 2.1—Percentages of homogeneous ability classes, by school SES

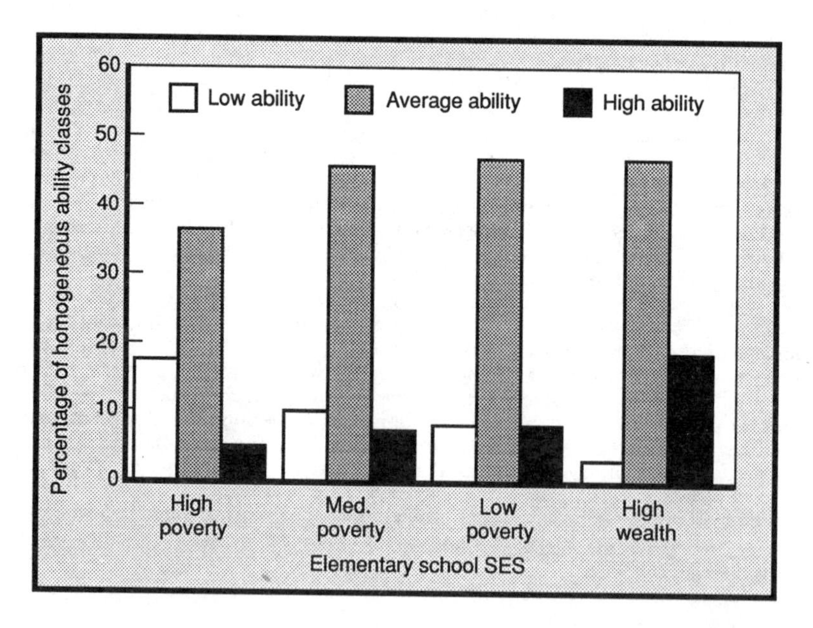

Fig. 2.2—Percentages of low-, average-, and high-ability classes in elementary schools, by school SES ($F = 11.05$, $P < 0.001$)

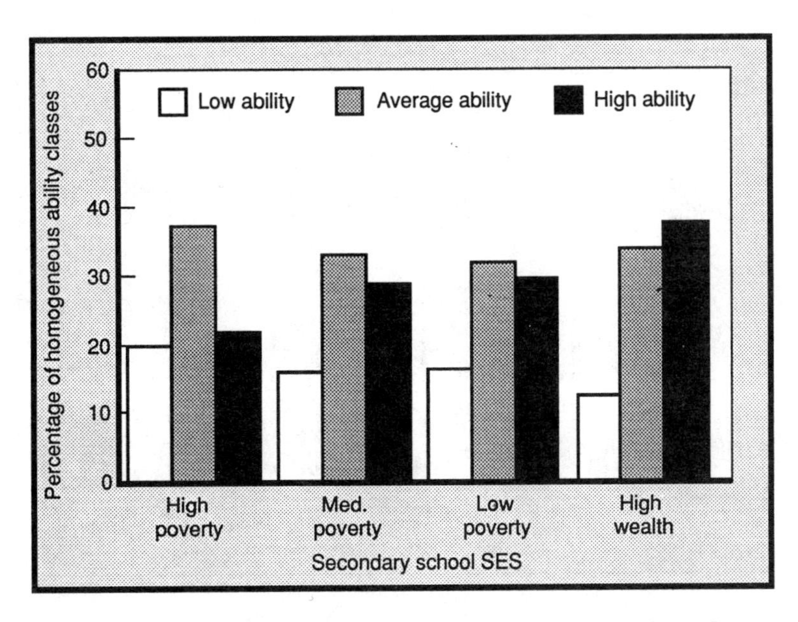

Fig. 2.3—Percentages of low-, average-, and high-ability classes in secondary schools, by school SES ($F = 3.92$, $P < 0.01$)

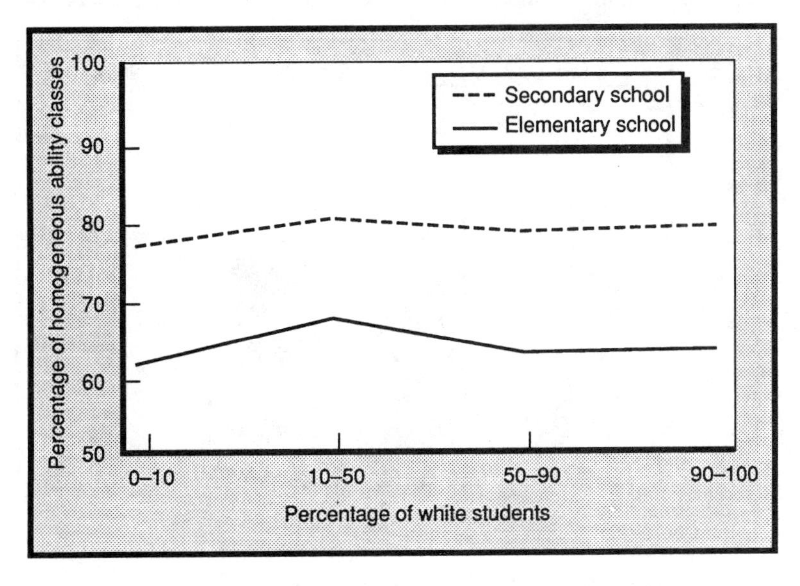

Fig. 2.4—Percentages of homogeneous-ability classes, by school racial composition

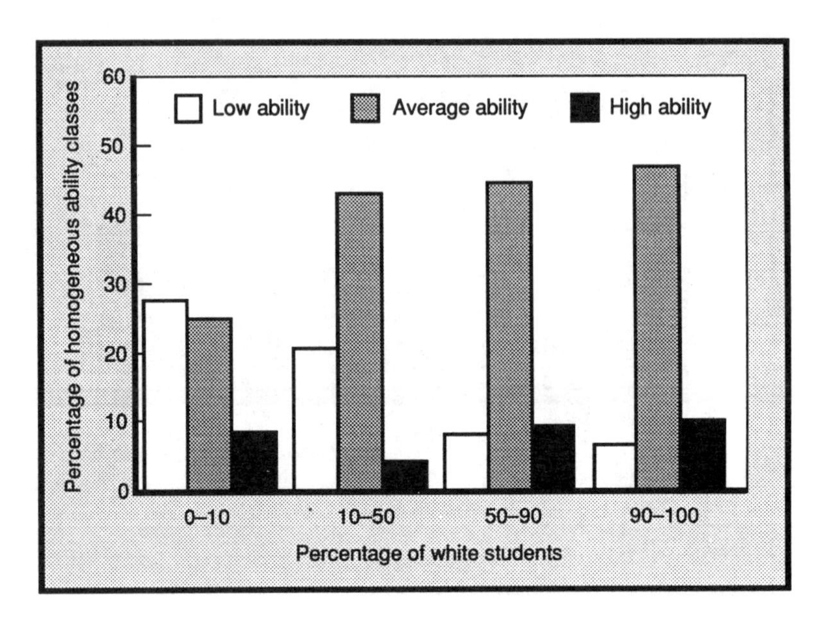

Fig. 2.5—Percentages of low-, average-, and high-ability classes in elementary schools, by school racial composition ($F = 10.34$, $P < 0.001$)

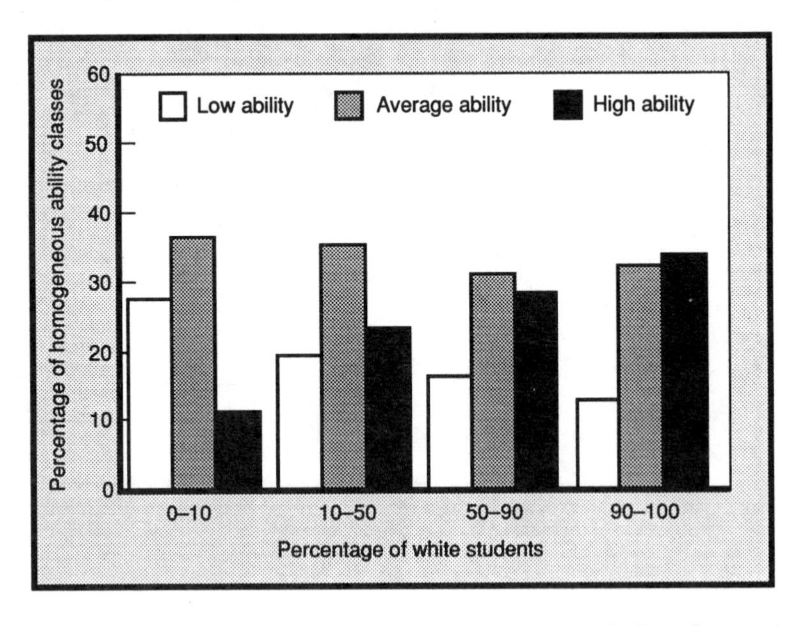

Fig. 2.6—Percentages of low-, average-, and high-ability classes in secondary schools, by school racial composition ($F = 10.70$, $P < 0.001$)

differences in the percentages of mixed or homogeneously grouped classes across types of schools. However, as Figs. 2.5 and 2.6 make clear, the types of ability-grouped classes differ significantly across schools with different racial compositions. Students who attend high-minority elementary and secondary schools are far more likely than students attending other schools to be enrolled in low-track science and mathematics classes and far less likely to be in high-track classes. Elementary school students' chances of being viewed as either average- or high-ability increase somewhat with the proportion of white students, the largest concentration of high-track classes occurring at the all-white schools. At secondary schools, the proportion of high-track classes increases dramatically as the proportion of white students increases, and there is a corresponding drop in the proportion of low-track classes.

There is also a second type of association between minority status and tracking. Tables 2.2 and 2.3 show the distribution of classroom ability classifications across classrooms with greater, roughly equal, and lower proportions of minority students than the minority enrollment at the school as a whole. We can consider these classes to be overrepresenting minorities, representative of the student population, or underrepresenting minorities.[10]

The distribution of ability-grouped classes at racially mixed elementary schools exhibited a highly significant pattern across the racial-composition categories.[11] Although disproportionately white classes were found to be about equally likely to be identified as low- or high-ability, disproportionately minority classes were *seven times more likely to be identified as low-ability than as high-ability.*

The association is even more dramatic in secondary schools: Not only were disproportionately minority secondary school classes far more likely to be judged low-ability (two-thirds of all such classes), disproportionately white classes were rarely judged this way. Differences in judgments about high ability were equally dramatic:

---

[10]For the school-level analysis, we used all secondary and middle schools in the data set. But because of the preponderance of 90 percent or more white schools in the sample, the classroom-level analysis was conducted for classes in schools with mixed racial composition (i.e., categories 2 and 3 of the school percentage-white variable). This analysis allows us to control for the racial composition of the schools when we consider the relationship between minority enrollment and ability classification at the classroom level.

[11]The tests of significance were conducted using class weights. For the school-level analysis, teachers' responses were averaged by school, and their weights were summed for each school, yielding a single observation. However, the data displayed in table form represent the teachers' responses directly.

Table 2.2

ABILITY LEVELS OF CLASSES IN ELEMENTARY SCHOOLS, BY RACIAL COMPOSITION OF CLASS RELATIVE TO SCHOOL ENROLLMENT

(In percent)

Class Enrollment[a] Relative to School Enrollment	Class Ability Level		
	Low	Average	High
Fewer minorities	22	52	26
Same (within 10%)	14	74	12
More minorities	30	66	4

NOTE: Table includes classes at mixed-race elementary schools (10 to 90 percent minority) only.

[a]Effect of class racial composition (relative to school's) is significant at the 0.001 level (F = 8.54).

Table 2.3

ABILITY LEVELS OF CLASSES IN SECONDARY SCHOOLS, BY RACIAL COMPOSITION OF CLASS RELATIVE TO SCHOOL ENROLLMENT

(In percent)

Class Enrollment[a] Relative to School Enrollment	Class Ability Levels		
	Low	Average	High
Fewer minorities	5	38	57
Same (within 10%)	21	50	29
More minorities	66	25	9

NOTE: Table includes classes at mixed-race elementary schools (10 to 90 percent minority) only.

[a]Effect of class racial composition (relative to school's) is significant at the 0.001 level (F = 143.45).

More than half of the disproportionately white classes were judged high-ability, compared with fewer than 10 percent of the disproportionately minority classes.

Our analyses of schools with different racial compositions and of classes at racially mixed schools show that African-American and Hispanic minorities face two potential barriers to science and mathematics opportunities. First, their access to high-track science and mathematics classes diminishes as the minority enrollment at their schools increases. Second, those who attend racially mixed schools are more likely than their white peers to be placed in low-track classes. There is thus a "double jeopardy" effect for African-American or Hispanic students. These relationships suggest strongly that minority students will suffer whatever disadvantages accrue to students in low-ability classes.[12]

Tracking is commonly viewed as a neutral, educationally sound response to a wide range of student aptitude and achievements, but the evidence we offer here confirms earlier work showing that such groupings are easily and commonly confounded with race and social class. Moreover, much prior research suggests that the differences in opportunities provided to ability-grouped classes *limit* instruction, rather than fine-tune it in ways that accommodate individual differences and promote learning (Oakes, Gamoran, and Page, in press). We examined the distribution of opportunities by the track level of science and mathematics classes to further investigate the possibility that in their efforts to accommodate differences in ability with different educational experiences, schools actually limit some students' opportunities to learn.

Because assessments of low academic ability and placements in nonacademic programs occur more frequently among low-income and minority students, the combined effects of ability and other background characteristics are of special importance in attempts to uncover factors related to underachievement and low participation among these groups. The recurrence of this relationship at both the school and classroom levels underscores the importance of considering ability grouping in any study of the relationship of students' characteristics to the distribution of opportunity.

---

[12]Because of the overlaps between minority status and poverty, a similar double disadvantage probably exists for low-income students. We were unable to test this hypothesis, however, because the NSSME survey collected no SES data at the classroom level.

III. ACCESS TO SCIENCE AND MATHEMATICS PROGRAMS

Obviously, the mathematics and science curriculum that is taught makes a difference in what students learn: When teachers teach particular mathematics and science topics, concepts, processes, and skills, students are more likely to learn them (Crosswhite et al., 1985; Husen, 1967; McKnight et al., 1987; Wolf, 1977). In this section, we examine the effects of school and classroom characteristics on the mathematics and science knowledge various groups of students have an opportunity to learn.

At the elementary level, the most basic—albeit limited—indicator of students' access to science and mathematics knowledge is the amount of time teachers spend teaching these subjects. An equivalent indicator at the secondary level is the number of science and mathematics courses schools offer. But to really understand the type of knowledge that students have an opportunity to learn, we also need to know the curricular goals that guide science and mathematics instruction for different students; the topics and skills being taught and the depth of course coverage; the numbers of topics covered; the scientific accuracy of the content of science and mathematics lessons; and the suitability of the lessons to students' developing cognitive abilities. There is little reliable data available, since large-scale research on these issues is difficult and costly.

In this section, we present new evidence about the quantity and quality of the science and mathematics curriculum that different groups of students experience in schools and classrooms and place this evidence in the context of findings from other research. First, we look at the extent of programs, measured by the amount of time spent on science and mathematics in elementary schools and *how many courses* in these subjects are offered in secondary schools. Then we consider the depth and rigor of the content in secondary school programs, as measured by the types of courses schools offer. We attempt to show the similarities and the differences in students' access to science and mathematics knowledge, and we examine the relationship of students' race, social class, community, and ability status to differences in their curricula.

TIME SPENT ON SCIENCE AND MATHEMATICS IN ELEMENTARY SCHOOLS

As noted above, the amount of class time that elementary teachers devote to teaching science and mathematics is the most basic indicator of students' access to these subjects, but it indicates little about the time students spend engaged in lessons or about the pace, content, and quality of instruction. Nevertheless, there is reason to believe that the amount of time spent studying a subject does influence the amount of learning that takes place.[1] This relationship is strongest in mathematics and science, since these subjects, unlike reading and social studies, are rarely learned in informal settings (Husen, 1967).

The time devoted to various subject areas varies considerably among schools. Goodlad (1984) surveyed teachers in thirteen elementary schools and found that their reported time spent teaching mathematics ranged from 45 minutes to 66 minutes per day. On average, mathematics occupied 20 percent of the total instructional time in these schools. Science typically occupied less time, but the variation among the schools was greater, with reported time ranging from 16 to 64 minutes. The average time devoted to science instruction was 28 minutes, about 10 percent of the total instructional time in these schools.

The NSSME data yield similar findings. Teachers of self-contained classes in grades K–3 reported spending an average of 43 minutes per day on mathematics instruction and 18 minutes on science.[2] Their counterparts teaching grades 4–6 reported spending 52 minutes on mathematics and 29 minutes on science. At both levels, the variations in science were greater than those in mathematics. Schools using "specialists" to teach science and/or mathematics typically devoted more time to science and slightly less time to mathematics (Weiss, 1987).

However, the variation in mathematics instructional time was not random. Schools and classrooms serving different groups of students spent significantly different amounts of time on mathematics lessons. Figure 3.1 shows that the number of minutes per day spent in math-

---

[1]See Carey (1989) for a review of the literature on this topic.

[2]Teachers reported the number of days per week they typically presented lessons in science and mathematics and the approximate number of minutes they spent in an average lesson. To obtain an estimate of the average number of minutes per day in each subject, we multiplied the reported minutes per lesson by the number of lessons per week and divided the result by 5.

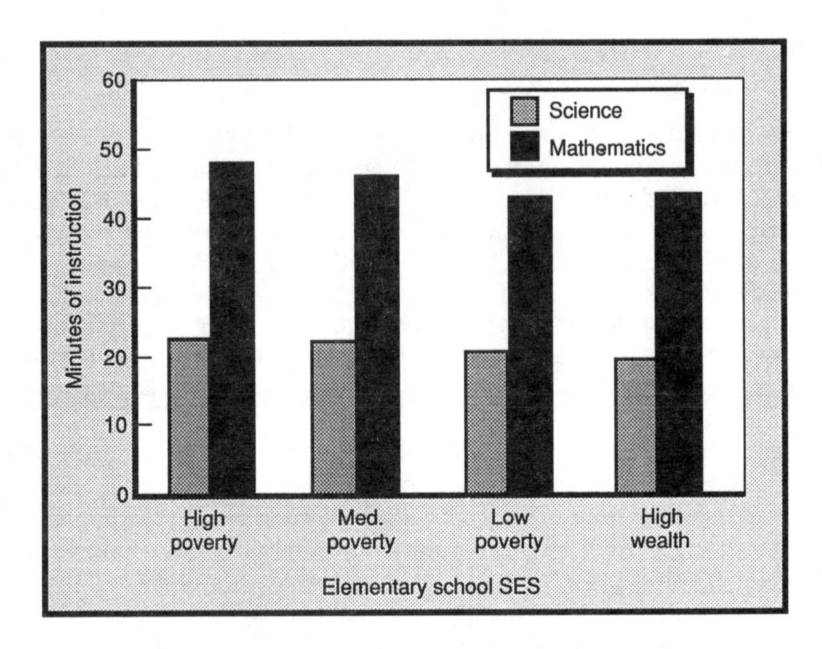

Fig. 3.1—Time spent on science and mathematics in elementary schools serving different student populations (for science, no significant differences; for math, $F = 4.92$, $P < 0.01$)

ematics instruction at the NSSME elementary schools differed by the percentage of low-income students enrolled.

Slightly more time per day was spent on mathematics at schools enrolling large concentrations of low-income students than at other types of schools: 50 minutes, on average, compared with 47, 43, and 44 minutes at schools serving increasingly advantaged populations. Similar time differences occurred among schools serving different racial groups and in different types of communities. Schools with the largest percentages of African-American and Hispanic students and those in inner cities devoted the most time to mathematics. Interestingly, the small amount of additional time in mathematics in low-SES, high-minority, and inner-city schools did not seem to be gained at the expense of science instruction, for which time allotments were similar across school types. Because these schools tend to have longer school days, they can accommodate longer periods of mathematics instruction. For example, the low-SES schools averaged 171 instructional minutes per day, and the highest-SES schools averaged

156 minutes. The percentage of the school day spent on mathematics was a relatively constant 28 percent across schools.

On its face, it is not surprising that teachers at low-SES, high-minority, and inner-city schools would spend more time on mathematics, since most of these schools qualify for federal assistance in the form of Chapter 1 funds targeted at improving disadvantaged students' basic skills (95 percent of the schools in our low-SES category reported that they qualified for these funds). However, federal assistance did not seem to affect directly the time spent on mathematics instruction. What it did affect was the time teachers said they spent on science. In all but the most affluent group of schools (only 15 percent of which qualified for federal funding), schools with Chapter 1 programs spent more time on science.[3] We speculate that raising students' mathematics achievement is thought to be important enough to command a large share of time *with or without Chapter 1 funding*. But the extra resources—in the form of specialist teachers and instructional materials—may enable participating schools to free up more teacher time and resources for science instruction.

In contrast to these school differences, the track level of elementary science and mathematics classes does not seem to affect the allocation of instructional time. Within NSSME schools of different types, students in low-, average-, and high-ability classes spent equivalent amounts of time on lessons. Because schools often regulate the minimum amount of class time for various subjects, teachers may have little discretion about how much time they direct students of different ability levels to spend on lessons.[4]

Thus, elementary schools with large concentrations of students who typically do poorly in mathematics seem to be attacking this problem by spending somewhat more time on mathematics lessons, presumably in hopes of giving disadvantaged and minority students a good start in mathematics. However, across all schools extra mathe-

---

[3] $F = 5.05, P < 0.05$.

[4] While we do not know with certainty that the homogeneous classes in our sample were groups of students who remained together for the whole day or groups of students pulled from heterogeneous homerooms for science and mathematics instruction, we suspect many were the former. We base our speculation on the fact that many elementary schools regroup students from heterogeneous classes for mathematics instruction, but few do so for science (Oakes, 1985; Slavin, 1987). However, since only slightly smaller percentages of science than mathematics classes were identified as homogeneous (63 percent of science and 70 percent of mathematics), we suspect that most of the homogeneous classes in the sample stayed together for most or all of the school day.

matics time is not being provided in those classes for the students who are perceived to be the least able.[5]

Our cross-sectional data do not permit us to determine whether the increased time allocations in schools with large concentrations of low-income and minority students are long-standing or reflect recent changes—perhaps in response to the press for additional time on basic subjects in recent school reform proposals (e.g., *A Nation at Risk*). Nor can these data indicate whether the extra mathematics time in disadvantaged and minority schools is spent in ways that are likely to help students overcome their historic patterns of low achievement. We know, for example, that the time teachers allocate to lessons is far less important than the time students are actually engaged in appropriate tasks (Berliner, 1979; Brophy and Good, 1986). Despite the limitations of these findings, however, achievement data showing a slow but steady decrease in the black-white mathematics gap in the elementary grades over the past decade suggest that some factor in basic skills instruction—perhaps even this small additional time allocation—may be having a modest benefit. However, other aspects of elementary school students' experiences, discussed in later sections, may offset this advantage.

SCIENCE AND MATHEMATICS PROGRAMS IN SECONDARY SCHOOLS

Secondary students' access to mathematics and science can be judged roughly by the courses in these subjects that schools make available. In this section, we are interested in (1) the extent of science and mathematics programs, (2) the content and rigor of the courses that make up those programs, and (3) the extent to which schools offer typical "gatekeeping" courses, i.e., courses that are usually prerequisites for participation in higher-level science and mathematics.

The Extent of Programs

Secondary schools vary considerably in the time they allocate to science and mathematics. Goodlad (1984) found that an average of 17

[5]Because the survey reported only how much time the classroom teacher spent on lessons, we do not know the extent of additional time low-track students might have spent in pull-out, remedial programs that supplemented regular classroom instruction.

percent of full-time teaching positions (equivalent to the percentage of instructional time) were devoted to mathematics in a sample of junior high schools. However, the range among the schools was 13 to 22 percent. In science, the average was 13 percent, but it ranged from 7 to 20 percent. Slightly less variability was found at the senior high school level, where, on average, 13 percent (with a range of 9 to 20 percent) of the teaching positions were allocated to mathematics, and 11 percent (with a range of 8 to 15 percent) were allocated to science.

What makes this variation worth noting is the connection between students' exposure to subjects and their achievement. Considerable evidence from national surveys (e.g., the National Assessment of Educational Progress (NAEP), National Longitudinal Study (NLS), and HSB) attests to the importance of high school coursetaking for mathematics and science achievement (Jones et al., 1986; Welch, Anderson, and Harris, 1982). Thus it is not surprising that the number of courses offered relates to what and how much students learn (Peng, Owings, and Fetters, 1981; Rock et al., 1985).

The extent of programs in secondary schools is a rough corollary of the time devoted to classes in elementary schools. Here, we are interested in the *numbers of class sections* offered in science and mathematics.[6] The numbers indicate *how much* science and mathematics is available and the extent to which students are taking advantage of available opportunities—either because of their own or their parents' wishes or because the school requires or presses students to enroll in science and mathematics courses. Of course, they reveal nothing about what *kinds* of courses are available. It is highly significant that differences in the extent of the programs at the NSSME secondary schools are not random, but rather are fairly consistently related to school population and community characteristics.[7]

[6]*Class sections* refers to the total number of classes offered by the school, not the number of discrete course offerings. For example, two schools may both offer 3 science courses: biology, chemistry, and physics. However, the first school may offer 3 sections of biology, 2 of chemistry, and 1 of physics, while the other offers 6, 5, and 2, respectively. While both schools offer 3 *courses*, the first offers 6 sections and the second offers 13. The importance of this is that it suggests greater participation rates for individual students if overall school enrollments are equivalent.

[7]The secondary and middle schools surveyed in the NSSME included schools with a variety of grade spans (9–12, 10–12, 7–9, 6–9, etc.). In this study, we usually used subsets of these schools. The 9–12 and 10–12 schools as a group represent high schools, and schools with maximum grades 8 or 9 represent junior high schools. The high school subgroup comprised about 360 schools; the junior high subgroup, roughly 200 schools. In some cases, we had to leave off the sampling weights because the schools were originally assigned a "senior" or a "junior" weight, but not both. This became a problem, since some of the 9–12 schools were classified as junior highs (those in which only ninth grade teachers were sampled). Thus, for some of our school-level

Middle Schools and Junior High Schools. Figures 3.2 and 3.3 show the differences in the extent of the mathematics and science programs at the junior high schools in the NSSME sample by school SES and racial composition.[8]

These patterns are quite different from those related to time allocations in elementary schools. Students attending the junior high and middle schools with the largest concentrations of *low-income students* had access to considerably less-extensive programs in both science (approximately 2.5 classes per 100 students) and mathematics (approximately 3 classes per 100 students) than those attending the most affluent schools (slightly more than 4 classes per 100 students in each subject); differences in science exceeded those in mathematics. The findings are only slightly different when the schools' racial makeup is considered. Science programs at schools whose student populations are more than 90 percent white were found to be significantly more extensive (nearly 4 classes per 100 students) than those at schools with similar percentages of minorities (about 2.5 classes); no significant race-related differences were found in the size of mathematics programs.

These findings, while perhaps disappointing, are not surprising. Many junior high school and middle school students are required to take only one semester or one year of science; additional science courses are either electives or recommended for high-achieving students. What the findings suggest is that high-SES and predominantly white schools either have greater science requirements or offer greater numbers of optional science courses.

The finding that low-SES junior high schools offer fewer sections of mathematics raises more troubling questions. In nearly every state, junior high school and middle school students are required to take mathematics each year, so mathematics classes in low-SES schools may be consistently larger than those in other types of schools—large enough to reduce the relative number of mathematics sections offered. Our data, in fact, do show such a trend. The lowest-SES schools had significantly larger mathematics classes than did the most affluent schools (averaging 26 and 23 students per class, respectively).[9] This

---

analyses, many of the more inclusive senior high schools (e.g., 9–12 schools) would not have the appropriate weights.

[8]To account for the size of the school and its grade span, which might affect the number of courses offered per student, we used an analysis of covariance to control for the logarithm of number of students per grade and the minimum and maximum grades in a school.

[9]$P < 0.05$.

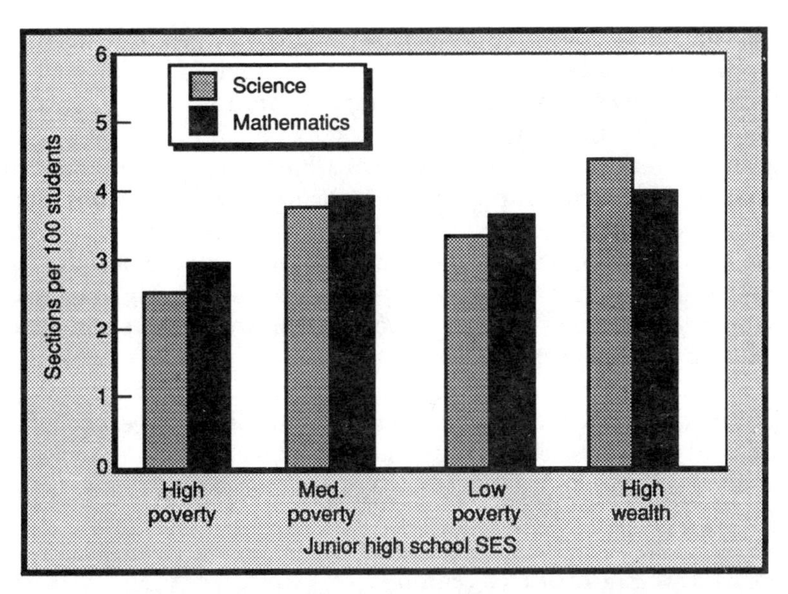

Fig. 3.2—Mathematics and science classes per 100 students in grade 6–9 junior high schools, by school SES (for mathematics, $F = 4.39$, $P < 0.01$; for science, $F = 13.69$, $P < 0.001$)

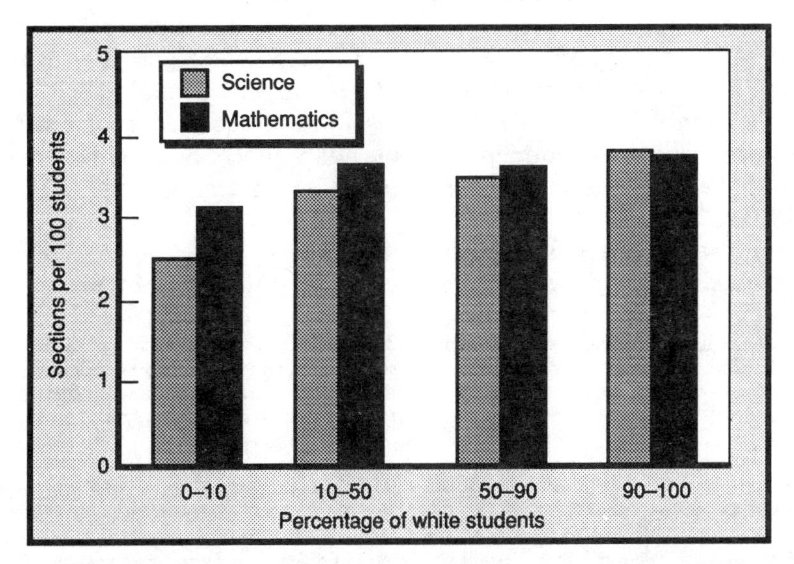

Fig. 3.3—Mathematics and science classes per 100 students in grade 6–9 junior high schools, by school racial composition (for mathematics, differences not significant; for science, $F = 3.64$, $P < 0.05$)

means that the junior high schools serving the largest concentration of low-income students allocated less time and fewer teachers to mathematics, even though all students were probably enrolled. Community type had little effect on the extent of junior high school programs. This suggests that low-SES schools in both inner-city and rural areas (the communities where most low-SES schools are found) offer students fewer science and mathematics opportunities.

These findings suggest that the possible *advantages* of more time on mathematics instruction in elementary schools serving low-income and minority students may disappear when students reach middle school or junior high school.

Senior High Schools. Science and mathematics programs are similar in size at most senior high schools, but high-SES schools have significantly larger science programs than other types of schools, and rural schools have the least extensive mathematics programs.[10]

Content and Rigor of Programs

Differences in the number of class sections reflect only the most general differences among programs. Data about the content and level of courses (e.g., advanced, general, or remedial) add considerable information, since programs at schools offering the same number of class sections may differ significantly in depth and rigor. For example, if a school offers no biology courses, no one at that school will have a chance to learn biology. These differences are important because the particular courses students take affect their achievement (Jones, 1984; Jones et al., 1987; Peng, Owings, and Fetters, 1981; Sells, 1982).

Because course offerings can constrain or encourage students' coursetaking in science and mathematics, differences in programs

[10]The subset of schools we used comprised all the schools with grade spans of 9–12 and 10–12. However, we considered only schools with more than 30 students in a grade to avoid the distortions extremely small schools can create. To account for school size and the presence or absence of a ninth grade (which can affect types of science offerings, for example), our model included the logarithm of number of students per grade and the minimum grade in a school. After controlling for these variables, we found that the distribution pattern by SES categories was significant at the 0.05 level for total science courses ($F = 3.27$, $P = 0.02$), but not for total mathematics courses ($F = 2.06$, $P = 0.11$). Racial composition (percentage white) did not result in significant differences with respect to total course offerings in either mathematics or science ($F = 1.22$, $P = 0.30$ for science; $F = 0.34$, $P = 0.80$ for math). Significant differences in mathematics programs were found in schools in different types of communities, however, with rural schools having the least-extensive programs ($F = 3.05$, $P < 0.01$).

may contribute to the achievement gaps among different student groups. HSB data show that senior high schools serving predominantly poor students typically offer fewer advanced placement (AP) courses (Ekstrom, Goertz, and Rock, 1988) and enroll proportionately fewer students in those they offer (Jones, Burton, and Davenport, 1984), and that students in college-preparatory programs at low-SES schools (the schools most minorities attend) typically take fewer academic courses (including mathematics and science courses) than their college-bound peers attending more-advantaged schools (Rock et al., 1985). These lower rates of academic coursetaking are linked to African-American students' lower levels of achievement in these subjects (Jones, 1984; Pallas and Alexander, 1983).

Figures 3.4 and 3.5 show the number of sections of general, college-preparatory, and advanced college-preparatory courses at the senior high schools in the sample.[11]

When the number of class sections is broken down by types of courses, the distribution patterns are significant or highly significant for all but general science classes. As the proportion of low-income and minority students at a school increases, the relative proportion of college-preparatory or advanced course sections decreases. Table 3.1 shows the significance of these differences.

The differences in the level of the classes offered cannot be accounted for by class size differences, as was the case among the junior highs. In fact, some of the differences in the sizes of various categories of classes in different types of schools compound the difference in opportunities among schools. For example, while no overall SES-related differences were found in the number of students in science classes, the size of college-preparatory and advanced college-preparatory mathematics classes increased with the affluence of the school.[12] This means that relatively more students were taking these courses at highly affluent schools than simple comparisons of the numbers of sections would indicate.

The Availability of Gatekeeping Courses

A third important dimension of students' access to science and mathematics knowledge is the extent of critical "gatekeeping" courses

---

[11]The courses that were classified in each of these categories are listed in the Appendix.

[12]$F = 3.03$, $P < 0.05$; and $F = 3.63$, $P < 0.05$, respectively.

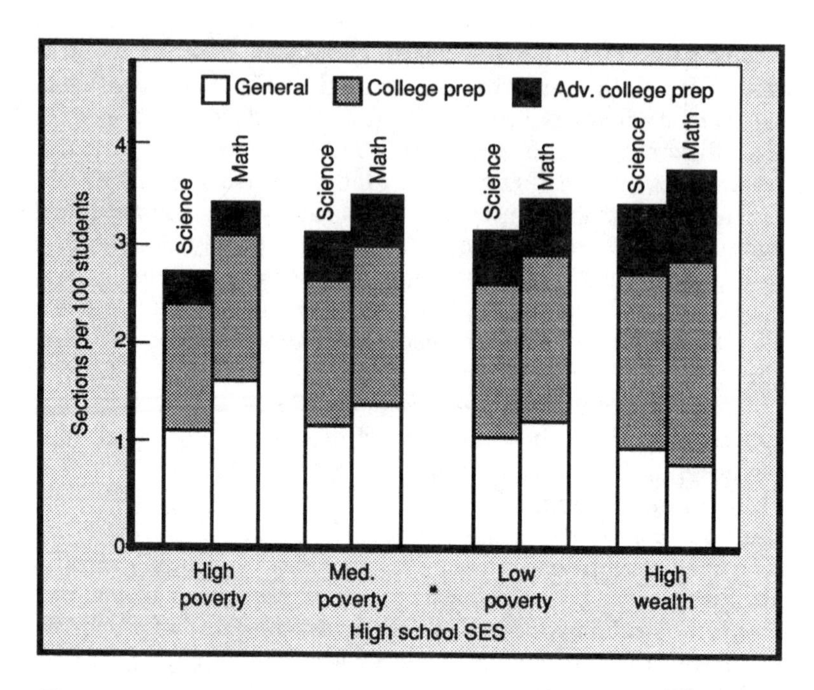

Fig. 3.4—Mathematics and science classes per 100 students in senior high schools, by school SES

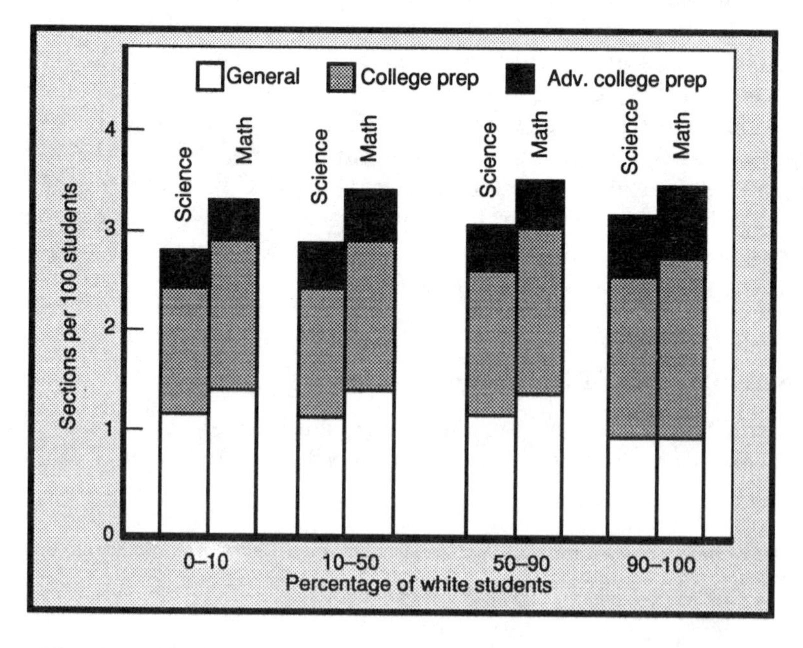

Fig. 3.5—Mathematics and science classes per 100 students in senior high schools, by school racial composition

Table 3.1

SIGNIFICANCE OF SES, RACE, AND LOCALE DIFFERENCES FOR SENIOR HIGH SCHOOL COURSE OFFERINGS

Independent Variable	Science F	Science P <	Mathematics F	Mathematics P <
	General Classes			
SES	1.30	0.27	12.42	0.001
% White	2.02	0.11	7.80	0.001
Locale	1.51	0.21	3.71	0.02
	College Prep Classes			
SES	8.23	0.001	11.23	0.001
% White	5.48	0.001	3.00	0.03
Locale	3.45	0.02	7.32	0.001
	Advanced College Prep Classes			
SES	11.30	0.001	16.54	0.001
% White	6.97	0.001	6.67	0.001
Locale	4.08	0.01	10.49	0.001
	All Classes			
SES	3.27	0.02	2.06	0.11
% White	1.22	0.3	0.34	0.8
Locale	2.20	0.09	4.05	0.01

offered at their schools, that is, courses that are especially important in qualifying students for post-high-school work in science and mathematics. At the senior high school level, the most critical course is calculus, because it is a prerequisite for entry into most science-, mathematics-, and technology-related majors at college. Without high school calculus, students must take beginning calculus classes in college. In many cases, this can make obtaining a baccalaureate degree in a quantitative field in four years very difficult. At the junior high school level, eighth grade algebra and ninth grade geometry are critical gatekeepers, since students who take these courses early are on track by grade 12 without having to double up classes or take mathematics courses during the summer.[13]

[13]The subsample of schools examined for calculus offerings was identical to that in the general mathematics and science offerings analysis, but the accelerated-mathematics analysis is based on our "junior high" grouping. The count of accelerated mathematics classes was determined by the number of algebra classes offered (in

Junior High Schools. We performed parallel analyses for the availability of algebra for eighth graders and geometry for ninth graders in the junior high school sample and the availability of calculus at the senior high level. Figure 3.6 shows the percentages of junior high schools, by SES and race, that offered at least one section of accelerated mathematics.

Despite what appear to be meaningful differences in the percentages of junior high schools offering accelerated mathematics classes—far fewer of the sampled low-SES and predominantly minority schools offered these classes—the differences among all SES categories for accelerated mathematics offerings were not significant.[14] A difference approaching significance, however, was found between the lowest-

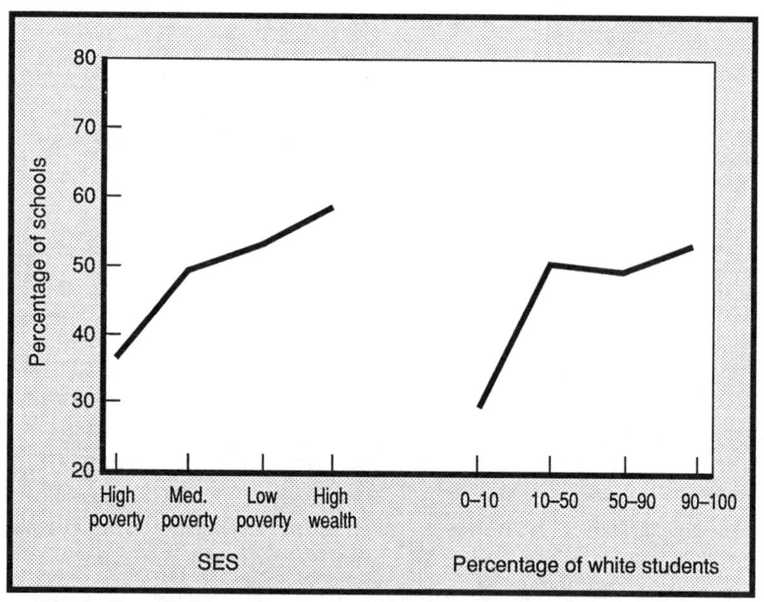

NOTE: Overall differences due to SES and racial composition were not significant, although groups 1 and 4 differed with P-values approaching significance (P = 0.086 for SES, P = 0.08 for racial composition).

Fig. 3.6—Junior high schools offering accelerated mathematics classes, by school SES and racial composition

---

schools whose final grade is 8), or by the number of geometry classes offered when algebra was also offered (in schools with a ninth grade).

[14] $F = 1.30$, $P = 0.27$.

and highest-SES schools.[15] Similarly, the differences across percentage white categories were not significant,[16] but again, the difference between high-minority and high-white schools approached significance.[17]

Figure 3.7 shows the numbers of sections of accelerated mathematics in schools of different types that offered these courses.[18] The distribution of the number of sections offered is significant across SES categories,[19] and it is highly significant across percentage-white categories.[20] Students attending high-SES or all-white schools have

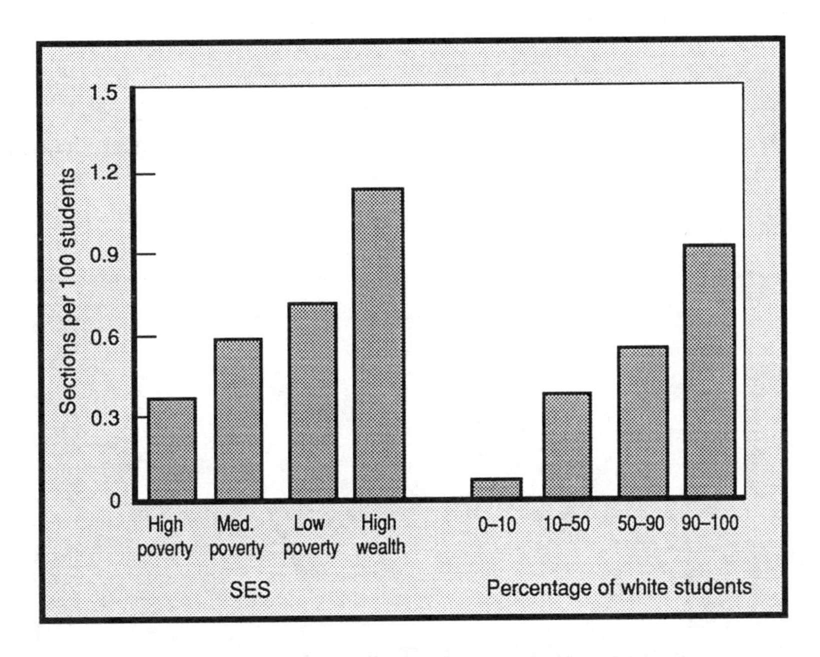

Fig. 3.7—Number of accelerated mathematics classes per 100 students in junior high schools offering accelerated mathematics classes, by school SES and racial composition (for SES, $F = 3.06$, $P < 0.05$; for racial composition, $F = 4.54$, $P < 0.01$)

---

[15] $P = 0.09$.

[16] $F = 1.04$, $P = 0.38$.

[17] $P = 0.08$. The lack of significance of these results may be attributable, in part, to the small number of schools in some of the categories.

[18] As in the earlier analyses, course offerings were divided by the average number of students per grade.

[19] $F = 3.06$, $P < 0.05$.

[20] $F = 4.54$, $P < 0.01$.

far greater opportunities to take critical gatekeeping courses that will prepare them for advanced mathematics and science courses in senior high school than students in low-SES and high-minority schools.

Senior High School. We were also interested in the percentage of senior high schools offering calculus classes.

Figure 3.8 shows the striking difference in students' opportunities to take calculus at schools of different SES levels. Predominantly white schools were found to be far more likely to offer calculus than high-minority schools,[21] although the differences across all four racial groupings did not reach statistical significance. Eighty percent of the predominantly white schools offered at least one section of calculus, whereas only about 50 percent of the high-minority schools did. This means that students at low-SES or high-minority schools who are prepared to take calculus in high school have far less opportunity to do so than their peers at economically more advantaged or predominantly white schools.

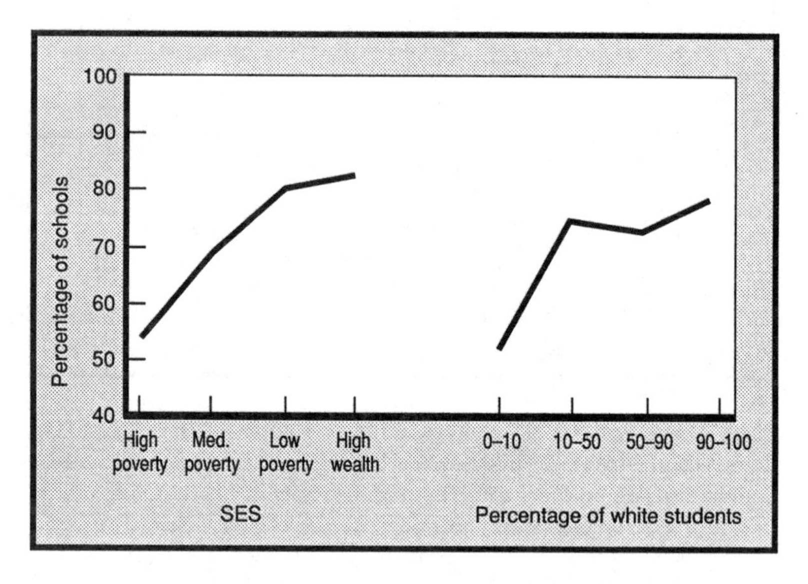

Fig. 3.8—High schools offering calculus classes, by school SES and racial composition (for SES, $F = 4.21$, $P < 0.01$; racial composition had no significant effect overall, although groups 1 and 4 differed with $P < 0.05$)

[21] $P > 0.05$.

Among schools offering calculus, a further dimension of opportunity is reflected in the number of sections offered relative to the size of the student body in various types of schools. Figure 3.9 shows that the opportunity to take calculus increases considerably as the proportion of low-income students drops, and that while mixed schools have relatively comparable numbers of calculus offerings, high-minority schools that offer calculus have far fewer sections. Moreover, the differences in students' access to calculus are minimized in these figures by the omission of schools that offered *no* sections.

It is difficult to determine exactly why schools offer the types of courses they do. Schools that offer few rigorous courses often do so because they have few students who are "qualified" to take those courses, that is, who meet test score or other criteria conventionally seen as prerequisites for learning content such as algebra, geometry, advanced mathematics, or mathematics-related science subjects. In fact, the most widely accepted explanation is that secondary schools'

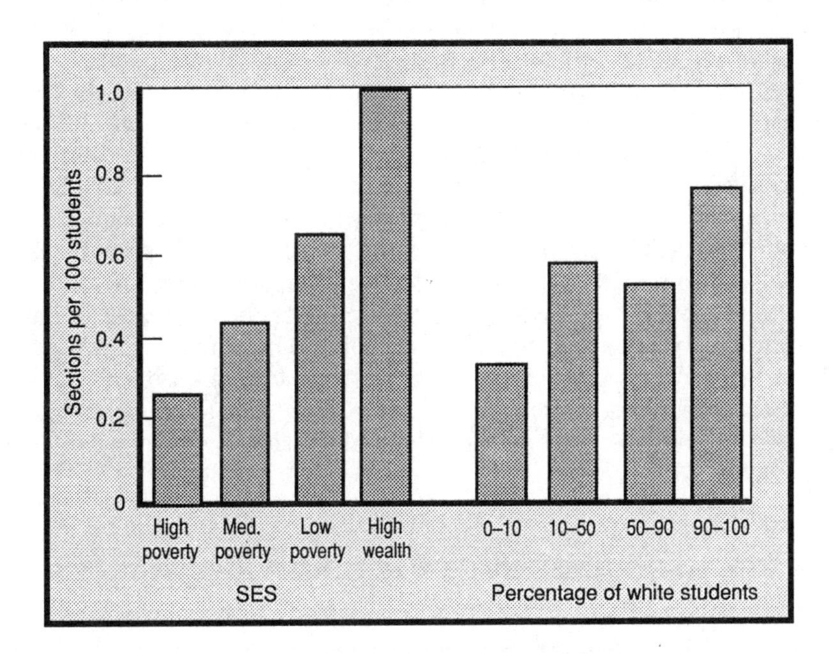

Fig. 3.9—Number of calculus classes per 100 students, by school SES and racial composition (for SES, $F = 11.28$, $P < 0.001$; for racial composition, $F = 3.05$, $P < 0.05$)

programs are constrained by earlier conditions that fail to develop the skills of disadvantaged students. But it should be noted that some schools bend conventional placement criteria and encourage lower-scoring students to take rigorous science and mathematics courses.

Some schools may simply place less emphasis on academic courses or, alternatively, more emphasis on more practical offerings, in the belief that such courses better meet their students' needs. Still other schools may lack staff who are able to teach certain courses, and this may be a significant factor in the paucity of curricular offerings in high-minority and high-poverty schools. Whatever the reasons, secondary schools serving low-income, minority, and inner-city populations offer fewer sections of college-preparatory or advanced courses and more general, applied, and remedial courses. The relatively small differences in the *numbers* of class sections offered in high school science and mathematics programs obscure substantial differences in the depth and rigor of programs at different types of schools.

Moreover, differences in students' opportunities to take critical gatekeeping courses affect their access to continuing study in science and mathematics. Of course, some qualified students, especially in cities, can transfer to schools that offer advanced programs. In these cases, students' opportunities may not be as restricted as our data suggest. However, this does not contradict the clear finding that fewer schools serving low-income and minority students offer the opportunity to begin the college-preparatory mathematics sequence in junior high school or to take calculus in senior high school. And among schools that offer these courses, the proportion of students who take them (as electives or as requirements) is far greater at high-income and predominantly white schools. These findings are quite disturbing: They are not simply consistent with the typically lower mathematics achievement of minority high school students, they strongly signal unequal access to valuable science and mathematics knowledge.

ACCESS TO COURSES WITHIN SCHOOLS

Schools' judgments about students' abilities and their likelihood of going to college affect access to various course offerings and shape the paths students take through the school curriculum. In most schools, fewer mathematics and science courses are available to low-track students or required for them (Guthrie and Leventhal, 1985; Sanders, Stone, and LaFollette, 1987; Vanfossen, Jones, and Spade, 1985).[22]

At the NSSME schools, traditional academic or standard college-preparatory courses and advanced college-preparatory courses are offered most often to students perceived as having high ability, somewhat less often to students thought to be of average ability, and hardly ever to students seen as having low ability (Table 3.2). Low-track students more often take general or applied science and mathematics courses.[23] While some of these courses include basic or applied "versions" of standard academic subjects (e.g., pre-algebra or introduction to algebra, general biology, applied chemistry, basic or "fun" physics), the content in most is quite different from the standard fare in academic science and mathematics.

Since low-SES and minority students are disproportionately enrolled in low-track classes and substantially underrepresented in high-ability classes, the likelihood of their being enrolled in academic and advanced courses decreases with the percentage of minority students in the class. Four-fifths of the predominantly white classes in the NSSME sample were taking traditional academic or advanced courses, whereas only 57 percent of the predominantly minority classes were taking such courses. Of course, these differences are

Table 3.2

DISTRIBUTION OF GENERAL, ACADEMIC, AND ADVANCED SCIENCE AND MATHEMATICS COURSES IN SENIOR HIGH SCHOOLS, BY ABILITY LEVEL OF CLASS

(Percentage of courses of each type taken)[a]

	Type of Course			
Class Ability Level	General	Academic	Advanced	Total
Predominantly low	74	25	1	100
Predominantly average	21	66	12	100
Predominantly high	5	40	55	100

[a]Chi-square = 581.521, $P < 0.001$.

[22]However, access to courses and coursetaking varies among schools with different tracking policies (Rosenbaum, 1986; Oakes, 1985). For example, students enrolled in nonacademic tracks in Catholic high schools typically take more academic courses and fewer electives than do their counterparts in public schools; the Catholic schools also place more students in academic tracks (Lee, 1986; Lee and Bryk, 1988).

[23]Business mathematics, consumer mathematics, remedial mathematics, general mathematics, technical mathematics, applied mathematics, basic or remedial science, general science, agriculture, current issues, medical technology, plant science, electronics, aviation, ecology, environmental science.

shaped in part by the fact that fewer academic courses are offered at high-minority schools.

However, similar differences appear when we control for racial composition of the entire school. As shown in Table 3.3, those classes in which white students are clustered tend to be academic, and those in which minority students are clustered tend not to be.

Just as differences in course offerings at schools of different types result in unequal access to science and mathematics content, curriculum differentiation through ability grouping leads to further race- and class-linked inequalities. Students judged to have low ability (disproportionately large numbers of minority students) have far less access to the standard academic subject matter in science and mathematics than do other students.

Table 3.3

DISTRIBUTION OF GENERAL, ACADEMIC, AND ADVANCED SCIENCE AND MATHEMATICS CLASSES, BY CLASS RACIAL COMPOSITION

(Percentage of courses of each type taken)[a]

	Type of Course			
Class Racial Makeup	General	Academic	Advanced	Total
Disproportionately minority	59	36	5	100
Racially proportionate (±10 percent)	23	51	25	100
Disproportionately white	15	49	36	100

NOTE: For these analyses, we considered only those schools with white populations greater that 10 percent, but less than 90 percent.

[a]Chi-square = 107.414, $P < 0.001$.

SUMMARY

This section has examined how students' access to science and mathematics knowledge varies with their race, social class, and neighborhood, and with the judgments educators make about their intellectual abilities. Except for slightly more time allocated for mathematics instruction in elementary schools with high concentrations of low-income and minority children, access to science and mathematics knowledge is limited for students from groups that consistently achieve and participate less in these areas. To the extent that they are enrolled in secondary schools where they are the majority, low-income students, African-Americans, and Hispanics have less-extensive and less-demanding science and mathematics programs available to them, and they have considerably fewer opportunities to take the critical gatekeeping courses that prepare them to pursue science and mathematics study after high school. This disadvantage is compounded by differences in students' opportunities within schools. Students who are thought to be of low ability are far less likely to be placed in traditional academic courses than are students judged to be more capable. These disparities undoubtedly reflect earlier and broader conditions that fail to develop the skills of large numbers of disadvantaged students rather than overt discrimination in the course enrollment process. But the net effect is that disadvantaged and minority students have considerably less access to knowledge that is considered necessary either for science and mathematics careers or for becoming scientifically literate, critically thinking citizens and productive members of an increasingly technological workforce.

Schools in the United States ration curriculum far more than those in many other countries. The Second International Math Study (SIMS) found that U.S. middle schools and junior high schools routinely sort eighth graders into four types of mathematics classes (remedial, typical, enriched, and algebra); this ability grouping was more extensive than that practiced in any other country studied. The SIMS also found that different types of classes provided students with quite different access to mathematics topics. For example, only the small percentage of U.S. students who were enrolled in algebra classes spent much of their class time studying algebra topics. In contrast, Japanese schools exposed nearly all their seventh graders to an intensive algebra curriculum. Only 13 percent of U.S. seventeen-year-olds were enrolled in advanced mathematics, and only about 20 percent of these advanced classes included calculus. This contrasts with many other countries where far greater percentages of students were enrolled in advanced mathematics (e.g., Hungary, Japan, Canada, Ireland, Scotland, Sweden, Thailand), nearly *all* of them in calculus courses (Travers and Westbury, 1989). Most surprising, the more exclusive populations of U.S. students who were taking calculus in the twelfth grade did not outscore the broader groups from other nations on advanced mathematics tests (McKnight et al., 1987). These data suggest that this highly selective system does not enhance achievement, even for those students enrolled in the most advanced courses.

IV. ACCESS TO QUALIFIED SCIENCE AND MATHEMATICS TEACHERS

Another factor in the lower achievement of low-income and minority students may be a lack of exposure to high-quality science and mathematics teachers. It has been widely believed, but not well documented, that predominantly minority and poor schools are less able to attract and retain qualified and experienced teachers. A recent report of the California Commission on the Teaching Profession argues, in fact, that disproportionate numbers of poor and minority students are taught during their entire school careers by the least-qualified teachers. The Commission report cites high levels of teacher turnover, larger numbers of misassigned teachers, and classrooms staffed by teachers holding only emergency credentials as problems in schools serving these "at risk" groups (California Commission on the Teaching Profession, 1985). National data from the mid-1980s indicate that teachers in inner-city schools are more likely to be uncertified than those who teach in the suburbs or rural areas (Darling-Hammond, 1985, 1987; Pascal, 1984). This problem has also been documented by anecdotes and some case-study work (Levy, 1970; Wise et al., 1987), but until now we have had little specific information about the distribution of science and mathematics teachers.

Schools enrolling large concentrations of low-SES students, African-Americans, or Hispanics are often perceived to be less desirable places in which to teach, and teacher shortages are likely to be felt most in these schools. Some national evidence already supports this contention. In 1983, there were about three times as many unfilled teaching vacancies (including positions that were withdrawn or for which a substitute was hired) in central cities as there were in other types of districts (NCES, 1985).

Teacher shortages are greatest in mathematics and science, especially in physical science. In 1981, more than half of the newly hired teachers in these fields either were not certified in science teaching or lacked certification in the specific courses they were to teach (NCES, 1983). Between 1972 and 1984, the number of newly graduating science and mathematics teachers decreased by 67 percent, and some estimates were indicating that as many as 30 percent of those teaching science and mathematics at the secondary level were unqualified or underqualified for their assignments (Johnston and Aldridge,

1984).[1] These data suggest that canceled courses, overcrowded classes, teaching misassignments, and the use of substitute teachers in mathematics and science courses are far more likely to occur in inner-city schools.

While little hard evidence is available to document the effects of teacher quality on students' achievement or choices, teachers are considered by almost everyone to be an important part of the educational process. Having well-prepared teachers who are knowledgeable in the subjects they teach is virtually a prerequisite to student learning (Darling-Hammond and Hudson, 1989). Thus, the teacher-quality gap among schools serving different student groups is in itself an important dimension of the distribution of opportunity to learn.

In this section, we first consider the extent of vacancies in the science and mathematics staffs of secondary schools of different types and the difficulties principals report in filling the vacancies that occur, based on NSSME data.[2] We next consider the principals' and teachers' perceptions of the competence of their mathematics and science staffs and the effects of teacher-related problems on the quality of mathematics and science instruction at their schools. We also consider the percentage of teachers employed at different types of schools who consider themselves to be "master" teachers in science and mathematics. Finally, we look at more tangible measures of teachers' backgrounds and qualifications: teaching experience, certification status, and academic preparation.

SHORTAGES OF QUALIFIED TEACHERS

Few schools have classrooms that are not staffed by teachers. When vacancies occur, school administrators do not usually hold positions open while they search for well-qualified replacements. If no qualified new hire can be immediately found, principals usually fill the opening with an unqualified teacher, use a substitute teacher, increase other teachers' class sizes or course loads, or cancel the course altogether.

The frequency with which vacancies occur and the difficulty principals have finding qualified teachers, then, provide useful information

[1]More recent data, however, suggest that most science courses are taught by teachers who specialize in science, although perhaps not in the specific subject taught (National Science Teachers Association, 1987; Weiss, 1987).

[2]Because most teachers in elementary schools teach all subjects, the issue of vacancies in the science and mathematics teaching staff is not relevant at that level.

about the quality of the teaching staff at different types of schools. And these data indicate that the teacher-shortage problem has the greatest effect on the access of poor and minority students to well-qualified teachers.

Where Do Vacancies Exist?

Secondary school principals in the NSSME sample did not vary significantly in their reporting of concern with mathematics vacancies, whereas science vacancies were of substantially greater concern in low-SES, high-minority, and inner-city schools. One plausible explanation for this unexpected finding (we would expect principals with the least-qualified teachers to express the greatest concern) is that principals in disadvantaged schools are less likely to perceive a mathematics position as "vacant" if an unqualified teacher can be found to fill it. In contrast, because science is often perceived as somewhat more specialized and rich in content than the low-level, computation-oriented mathematics classes that dominate the curriculum in low-income, minority schools, principals may be quicker to perceive a vacancy if a science course is taught by someone without science expertise.

Figure 4.1 shows the extent to which NSSME principals were concerned about filling biology or life science vacancies.[3] Principals at all types of schools expressed concern about vacancies in science, but those at high-poverty, high-minority, and inner-city schools expressed this concern most frequently: 97 percent of the inner-city secondary school principals indicated that science vacancies were a problem, compared with 64, 67, and 70 percent of the principals of schools in other types of communities. Of course, the extent of vacancies often reflects high turnover and/or inadequate staffing.

How Hard Are Vacancies to Fill?

The extent of the difficulty principals report having in filling vacancies with well-qualified teachers also varies. Again, principals

---

[3]Principals of secondary schools were asked, Does your school find it difficult to hire fully qualified teachers for vacancies in each of the following fields? They responded either *yes, no,* or *no vacancies/does not apply*. Separate responses were obtained for mathematics, several science subjects, and other school subjects. The percentages in Fig. 4.1 represent secondary school principals who answered either *yes* or *no,* since both answers indicate that staff vacancies are a problem.

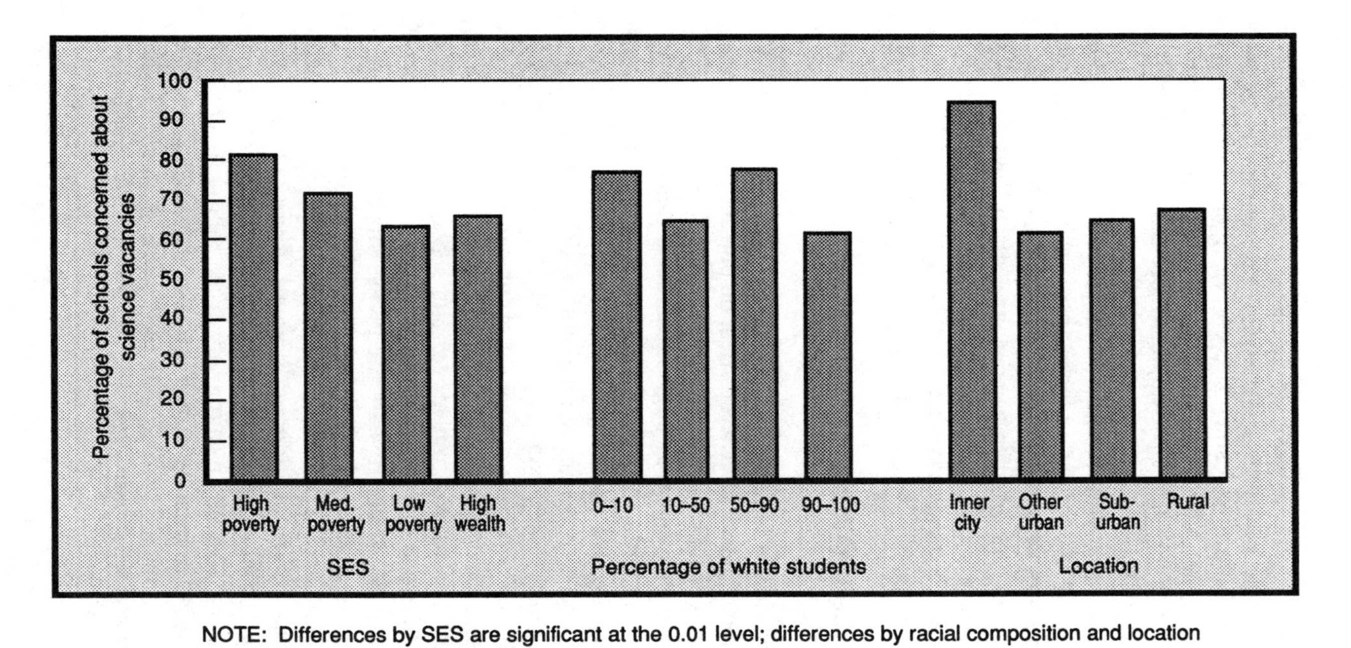

NOTE: Differences by SES are significant at the 0.01 level; differences by racial composition and location are significant at the 0.001 level.

Fig. 4.1—Percentages of secondary schools where life science/biology teacher vacancies were of concern to principals, by school SES, racial composition, and location

at schools with the least-advantaged students, the highest proportions of minority students, and inner-city locations have the greatest difficulty. Figures 4.2 and 4.3 show the percentages of principals in schools of various types who reported difficulty filling vacancies in mathematics and biology—the two subjects most commonly taught in all types of secondary schools.

These findings suggest that low-income and minority students are seen as less desirable students to teach, and inner-city communities are thought to be less desirable places for teachers to work in. In addition, salaries at inner-city schools are often lower than in surrounding areas, working conditions are poorer, and there are fewer material resources to work with.

WHICH SCHOOLS HAVE THE MOST-QUALIFIED TEACHERS?

It is difficult to define teacher quality, since good teaching results from a combination of many characteristics, no one set of which works best for all students or all classrooms.[4]

Nevertheless, we can obtain a rough measure of teacher quality by considering the judgments that principals and teachers make about teaching staff quality and the judgments teachers make about their own competence, and by examining teachers' credentials, educational background, and years of experience.

The distribution of teachers across different types of schools provides information about the overall level of human resources available to implement the curriculum at various schools and about the level of success different types of schools have in attracting teachers who are well-qualified. It also reveals the access various groups of the nation's students as a whole have to well-qualified teachers. The distribution of teachers in various types of classes also provides information on the effects of judgments about students' abilities on access to well-qualified teachers. Finally, this information may indicate patterns of teacher assignment that relate to the racial composition of classes in racially mixed schools.

[4]See Darling-Hammond and Hudson, 1989.

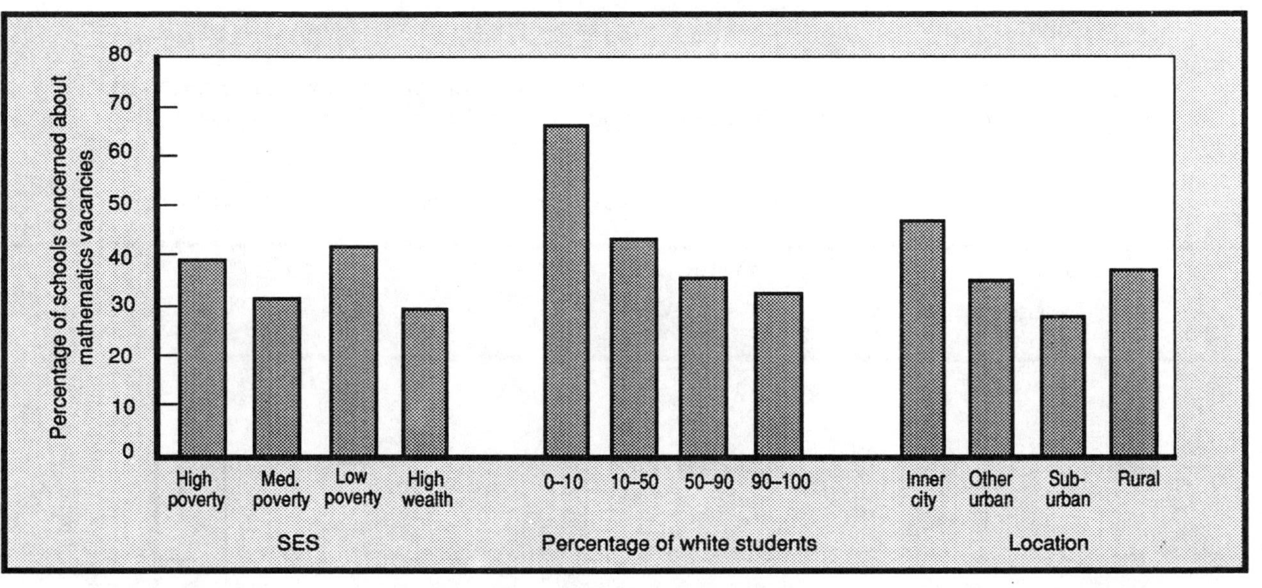

NOTE: Differences by SES are not significant; differences by racial composition are significant at the 0.01 level; differences by location are significant at the 0.05 level.

Fig. 4.2—Percentages of secondary schools where principals reported difficulty filling mathematics teacher vacancies, by school SES, racial composition, and location

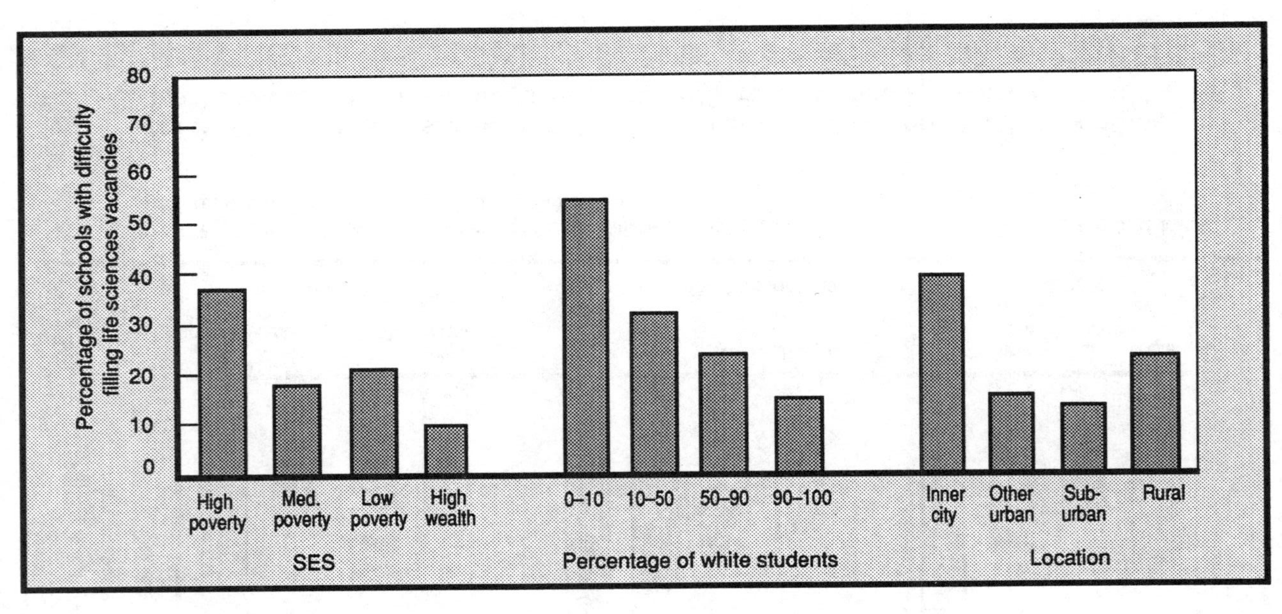

NOTE: Differences by SES, racial composition, and location are significant at the 0.001 level.

Fig. 4.3—Percentages of secondary schools where principals reported difficulty filling life science/biology teacher vacancies, by school SES, racial composition, and location

Perceptions of Teacher Competence

In the NSSME, secondary school principals were asked to indicate how many of the science and mathematics teachers at their schools they considered to be highly competent, competent, and not competent. While such judgments do not define quality, they do indicate the degree to which others perceive it. Figures 4.4 and 4.5 show the average percentages of teachers whose principals judge them highly competent at various types of schools.

Principals at schools with large concentrations of low-income or minority students and at inner-city schools reported that fewer of their teachers were highly competent. Principals at high-wealth, racially mixed, predominantly white, and suburban schools were far more satisfied with the quality of their science and mathematics teachers.

We also looked for patterns in the percentage of principals who reported that teacher-related problems affect the quality of science and mathematics instruction at their schools. The NSSME principals were given a list of possible factors that may cause serious problems in science and mathematics, and were asked to indicate whether a lack of teacher interest and/or inadequate preparation to teach these subjects caused serious problems at their schools. Only 8 percent of the more than 1,000 elementary and secondary principals who responded said that this was the case. Among this small percentage, no significant patterns of differences appeared across all school types at the elementary school level, although principals at the most affluent schools reported these teacher problems far less often than did those at other types of schools. At the secondary school level, however, patterns of teacher problems do seem to be related to the racial mix in the student body and the type of community in which schools are located. The social class composition of the schools bore no apparent relationship to principals' judgments that teacher interest and/or preparation caused serious problems for their science and mathematics programs (Fig. 4.6).

Teachers may be more aware of the instructional problems caused by disinterested or underprepared teachers.[5] Again, we see no pat-

[5]It is not possible to compare principals' and teachers' perceptions of problems with any certainty, since the NSSME questionnaires worded these questions differently for the two groups. While principals were asked to mark factors that cause "serious" problems, teachers were allowed a broader range of responses. For each factor—including lack of teacher interest and lack of teacher preparation—teachers were asked whether it was a "serious problem," "somewhat of a problem," or "not a significant problem."

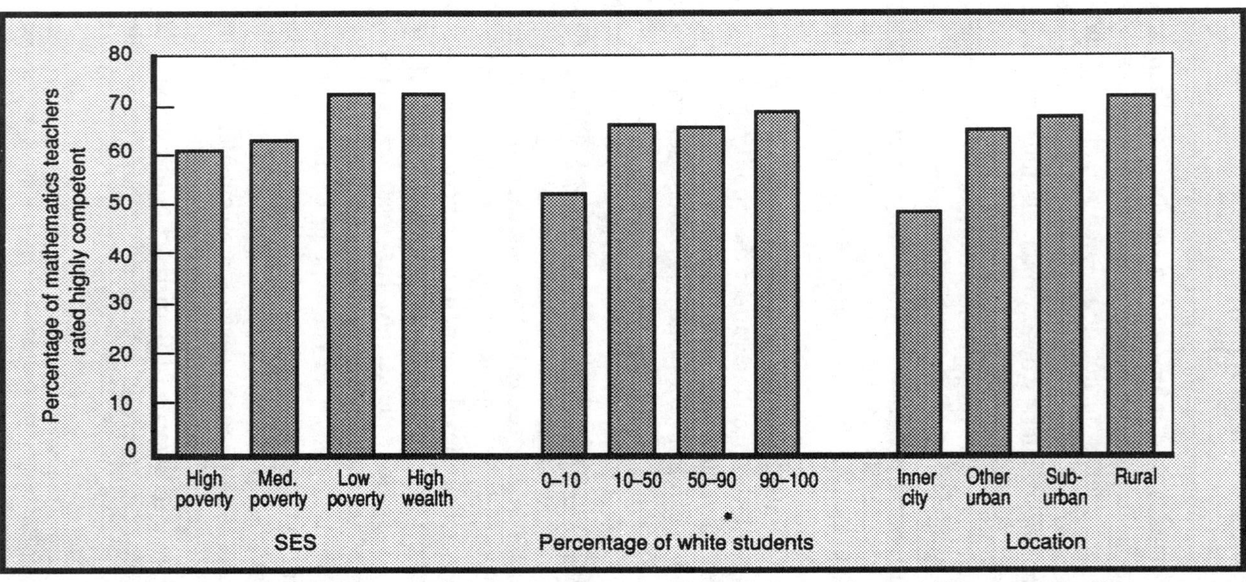

NOTE: Differences by SES are significant at the 0.01 level; differences by racial composition are not significant; differences by location are significant at the 0.001 level.

Fig. 4.4—Proportion of secondary school mathematics teachers considered highly competent by their principals, by school SES, racial composition, and location

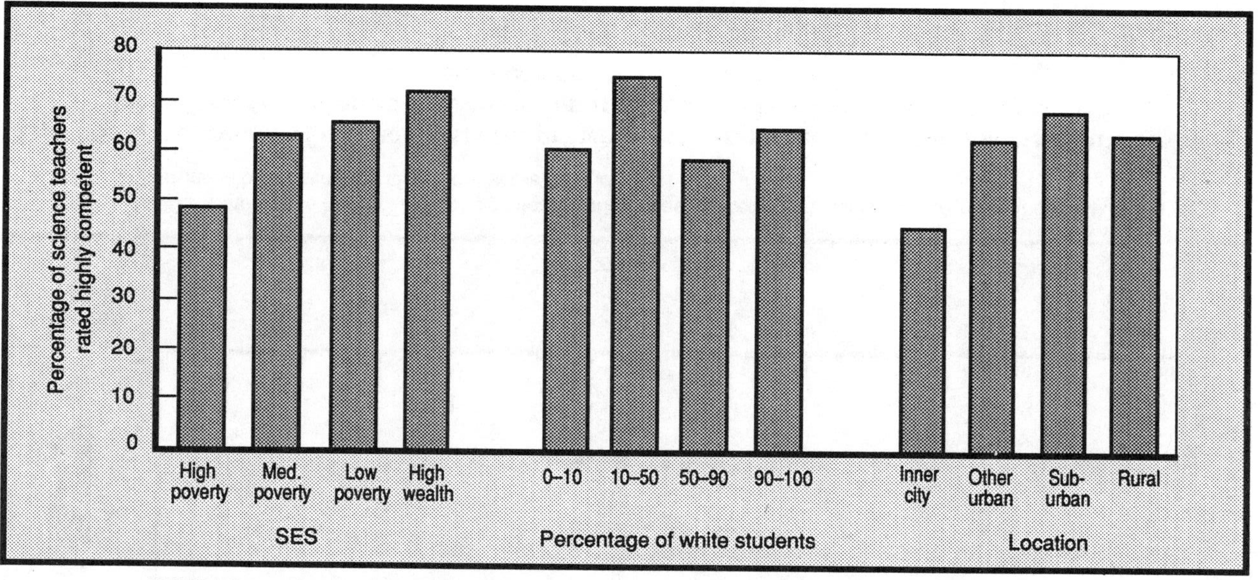

NOTE: Differences by SES and location are significant at the 0.001 level; differences by racial composition are significant at the 0.05 level.

Fig. 4.5—Proportion of secondary school science teachers considered highly competent by their principals, by school SES, racial composition, and location

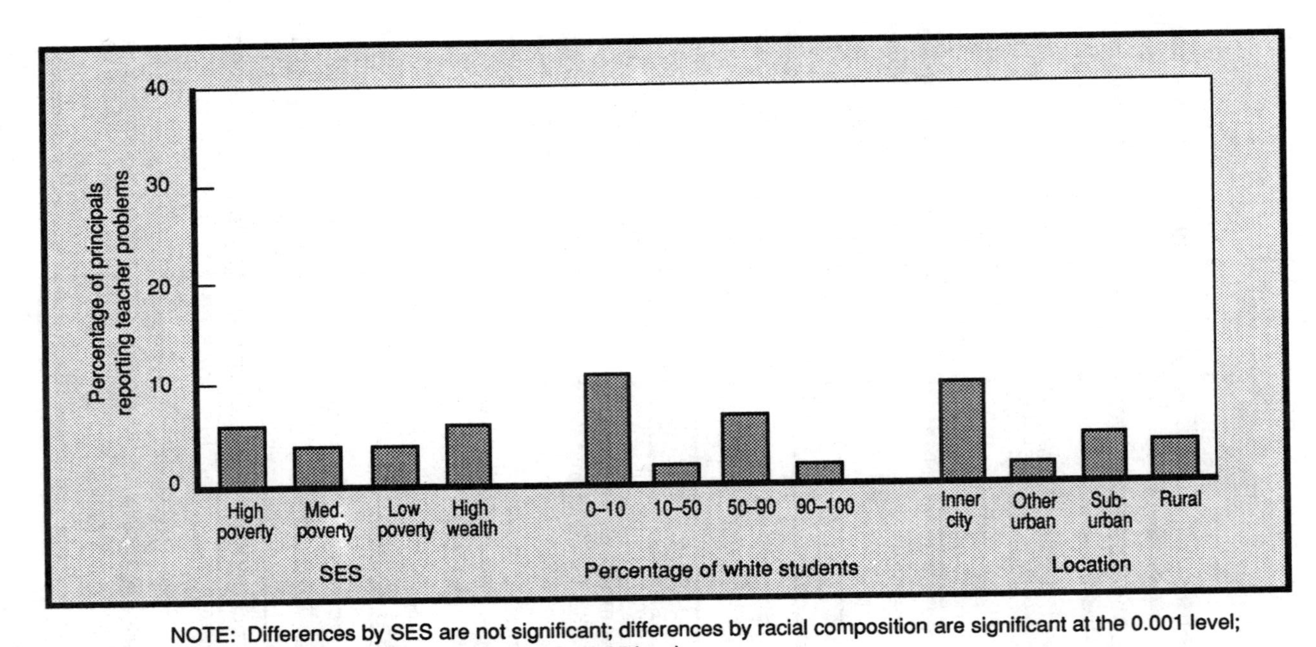

NOTE: Differences by SES are not significant; differences by racial composition are significant at the 0.001 level; differences by location are significant at the 0.05 level.

Fig. 4.6—Percentages of secondary school principals who reported serious problems resulting from lack of teacher interest or preparation in science and mathematics, by school SES, racial composition, and location

tern of differences at the elementary school level, but quite dramatic differences among secondary schools (Fig. 4.7).

Teachers at high-poverty, high-minority, and inner-city schools reported far more frequently than teachers at other types of schools that a lack of teacher interest or insufficient background posed problems for science and mathematics instruction. In each category of schools, teachers in the least-advantaged schools reported problems at least twice as often as those in other types of schools.

Finally, teachers were asked to indicate the degree to which they considered themselves "master" teachers in science or mathematics.[6] On the whole, elementary teachers seemed less confident about their science and mathematics teaching than their counterparts in secondary schools, especially in science. Only 12 percent of the elementary teachers reported that they were master science teachers, while 37 percent of those asked about mathematics said they were. In neither subject did teachers in schools of various types differ, on average, in their confidence levels. In contrast, the distribution of highly confident secondary teachers was not so even: More than half of them in both science and mathematics agreed that they were master teachers. But far fewer teachers in inner-city and rural schools and schools enrolling large concentrations of low-income children perceived themselves to be this competent.[7]

In sum, then, the perceptions of principals and teachers converge around the issue of teacher quality. Schools serving disadvantaged and minority students reported suffering from teacher shortages and teacher quality problems far more than other types of schools.

Teachers' Formal Qualifications

Elementary Schools. Teachers' formal qualifications follow many of the same distributional patterns as teacher shortages and perceptions of teacher competence. There is little evidence of differences in the certification status, academic backgrounds, and teaching experience of elementary teachers working in different types of

[6]Teachers were asked to respond to the statement, "I consider myself a 'master' science (or math) teacher," by checking either *strongly agree, agree, no opinion, disagree,* or *strongly disagree.*

[7]As the concentration of low-income students decreased at schools, the average level of perceived competence rose ($F = 4.98$, $P < 0.01$). Similarly, teachers at schools in urban (outside the inner city) and suburban schools were more self-confident ($F = 4.88$, $P < 0.01$).

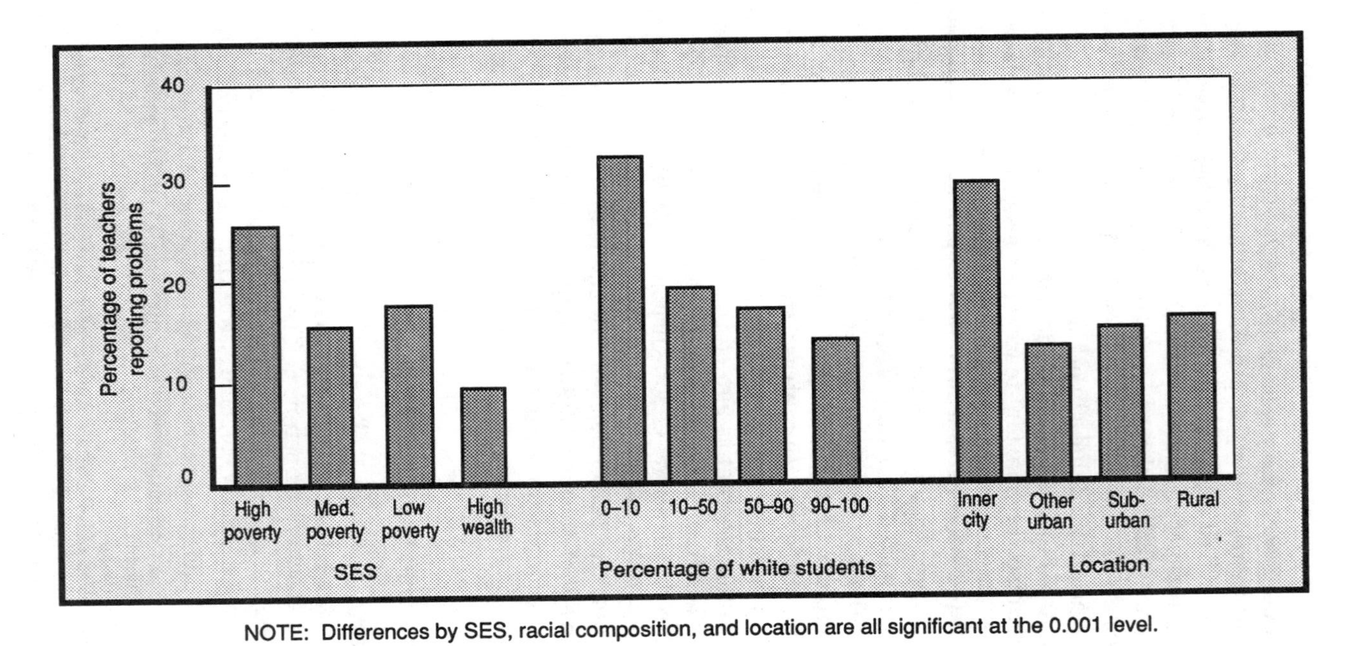

NOTE: Differences by SES, racial composition, and location are all significant at the 0.001 level.

Fig. 4.7—Percentages of secondary school teachers who reported serious problems resulting from lack of teacher interest or preparation in science and mathematics, by school SES, racial composition, and location

schools. However, there is one striking exception: While few elementary teachers were found to have a bachelor's degree in either mathematics or science or a degree in education with a mathematics or science emphasis (only 4 percent over the entire sample), teachers at elementary schools with the highest concentrations of minority students were more likely to have such degrees than were teachers at schools with other racial mixes (16 percent in high-minority schools; 3, 5, and 6 percent in the other categories based on racial composition).[8] The large percentage of minority teachers with degrees accounts for the overall higher level of teacher qualifications at the predominantly minority schools. These schools employ the largest percentages of minority teachers,[9] who are more likely to have mathematics and science degrees than their counterparts at other types of schools. At predominantly minority schools, minority teachers are also more likely than whites to have mathematics and science bachelor's degrees. The number of well-qualified minority teachers was large enough to counter the smaller proportions of degree-holding white teachers at these schools.

A major reason for the presence of well-qualified teachers at Chapter 1 schools is that these schools are more likely than others to have mathematics specialists. Many of these teachers are specially trained and certified in elementary mathematics education, and city schools were among the first to train and hire specialists. Also, many highly qualified minority teachers became committed to these schools years ago. Young, inexperienced white teachers who teach in high-turnover, inner-city schools are generally "paying their dues" before they can transfer out to other schools.

One interesting exception to the pattern of more highly qualified minorities being employed at high-minority schools was found in the all-white elementary schools. The few minority teachers at these schools had more science and mathematics degrees than minority teachers at any other type of school (22 percent held these degrees); however, their overall numbers were so small that their presence did not give the schools a relative advantage on this dimension.

Secondary Schools. In contrast to the similarities in teacher qualifications at elementary schools, the qualifications of secondary school teachers at schools of different types differ substantially. Teachers at schools with predominantly economically advantaged white students and suburban schools are, on average, more qualified. There are significant differences in the average amount of teaching

---

[8] $F = 3.64$, $P < 0.05$.

experience at different types of schools, but students attending predominantly white, high-SES, and suburban schools have greater access to well-qualified science and mathematics teachers.

Teachers at high-minority, high-poverty, inner-city schools are slightly less likely to have state certification in any subject.[9] And as shown in Figs. 4.8, 4.9, and 4.10, the differences among types of schools become more pronounced as the level of qualification increases beyond basic certification. Teachers in schools serving minority and disadvantaged students are less likely to be certified to teach science and mathematics or to hold bachelor's and/or master's degrees in these fields. They are also less likely to meet the standards of the National Science Teachers Association or the National Council of Teachers of Mathematics.

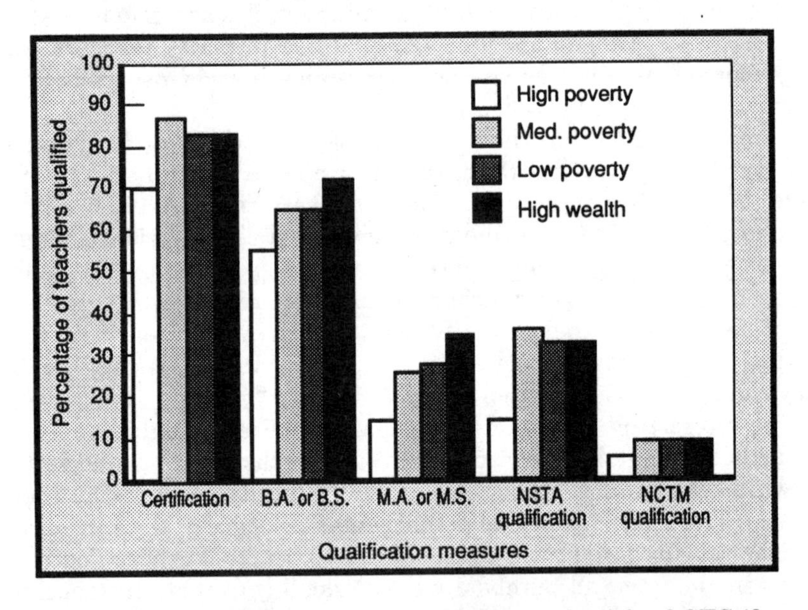

Fig. 4.8—Secondary teachers' qualifications, by school SES (for certification, master's degree, NSTA qualification, $P < 0.001$; for bachelor's degree, $P < 0.01$; differences for NCTM qualification were not significant)

[9]For racial composition, $F = 5.19$, $P < 0.01$; for locale, $F = 2.69$, $P < 0.05$. No significant differences were found related to schools' SES.

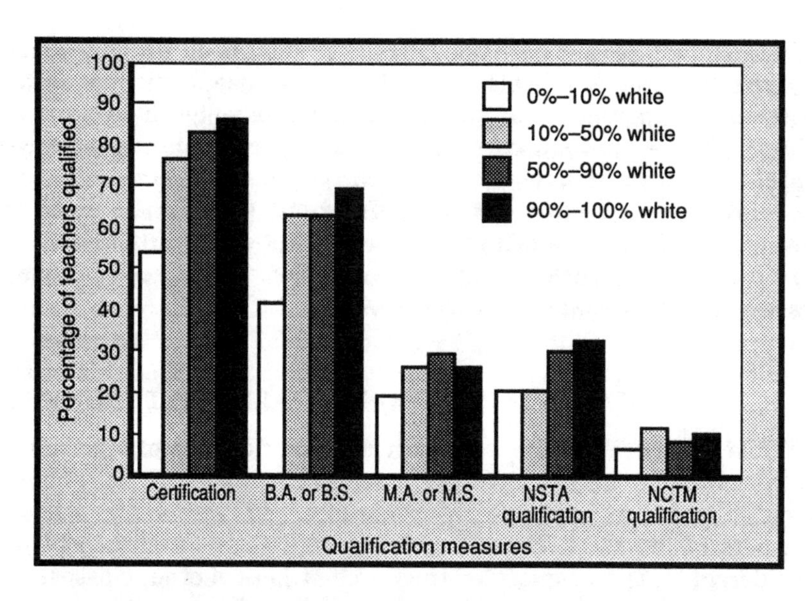

Fig. 4.9—Secondary teachers' qualifications, by school racial composition (for certification and bachelor's degree, $P < 0.001$; for NSTA qualification, $P < 0.05$; differences for master's degree and NCTM qualification were not significant)

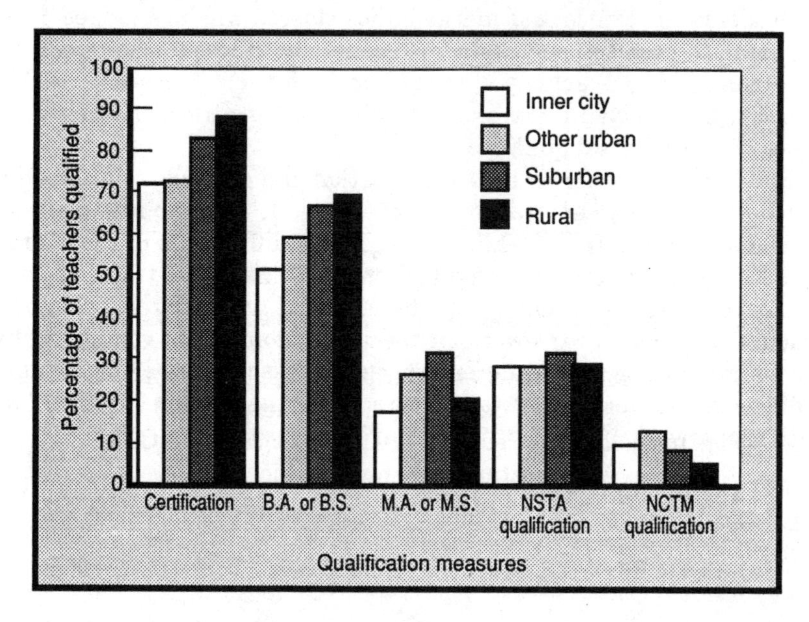

Fig. 4.10—Secondary teachers' qualifications, by school location (for certification, bachelor's degree, master's degree, $P < 0.001$; differences for NSTA and NCTM qualification were not significant)

Formal qualifications are not the only standards on which teachers at different types of schools vary. There are also significant differences in the amount of formal computer training teachers have received: Those at high-poverty schools,[10] predominantly minority schools,[11] and inner-city and rural schools[12] have received substantially less than those at other types of schools. Clearly, the small advantage in teaching quality enjoyed by students participating in mathematics programs in high-minority elementary schools disappears by junior or senior high school.

WHICH CLASSES HAVE THE MOST-QUALIFIED TEACHERS?

While there are few differences in the qualifications of science and mathematics teachers in elementary school classes at different ability levels, there are considerable differences in secondary schools. Teachers of low-track classes in junior and senior high school are considerably less well-qualified than are teachers of other classes.

The Relationship Between Judgments About Student Ability and Teacher Qualifications

Students in science and mathematics classes who are judged to be low in ability tend to be taught by teachers with less teaching experience than those teaching average- or high-track classes.[13] Additionally, as shown in Fig. 4.11, teachers of low-track classes rank lower on most formal qualifications.

In addition to differences in certification and degrees, relationships can be found between tracks and the amount of computer training teachers have received. While there are few differences in the overall amount of computer training (self-taught, inservice training, and coursework) experienced by teachers, there are differences in the amount of college coursework in the use of computers completed by teachers working with different types of classes. Teachers of high-ability classes were found to be more likely than others to have had such coursework.[14]

---

[10] $F = 4.37, P < 0.01$.
[11] $F = 3.35, P < 0.05$.
[12] $F = 9.28, P < 0.01$.
[13] $F = 5.47, P < 0.01$.
[14] $F = 6.40, P < 0.01$.

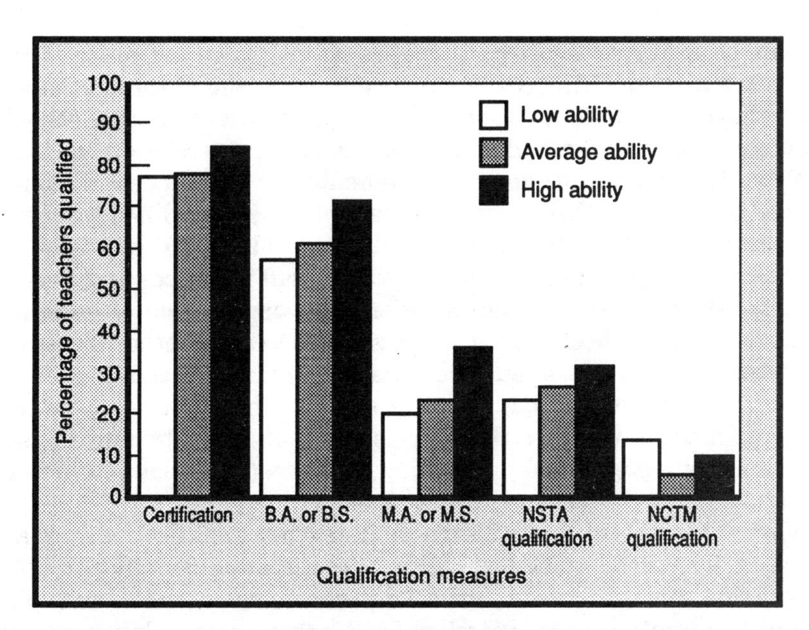

Fig. 4.11—Secondary teachers' qualifications, by ability level of class to which they are assigned (for certification, bachelor's degree, master's degree, $P < 0.001$; for NCTM qualification, $P < 0.01$; for NSTA qualification, $P < 0.05$)

Finally, we found significant differences in the extent to which science and mathematics teachers of different ability-level classes perceive themselves to be highly competent. Teachers of high-track classes felt most strongly that they were master teachers; teachers of average-track classes were somewhat less likely to characterize themselves as master teachers; and teachers of low-track classes had the poorest image of their abilities.[15]

The Effects of School Type and Class Type

The finding that secondary students in classes of different track levels differ in their access to well-qualified teachers must not be interpreted simplistically, since some class-level differences are confounded by large school-level differences in teacher qualifications. Low-SES, high-minority, and inner-city junior and senior high schools have, on average, the least-qualified teachers *and* have dispropor-

[15] $F = 27.92$, $P < 0.01$

tionate percentages of low-track classes (as discussed in Section II), so some track-level differences may result from the disproportionate percentages of students in low-ability classes who attend schools where access to well-qualified teachers is constrained by overall staff deficiencies. While this does not invalidate the finding that students judged to be low in ability are being taught by less-qualified teachers, it is important to determine the extent to which the matching of teachers and tracks is a function of school staff resources and the extent to which it results from a differential distribution of qualified teachers within schools. That is, we need to know whether low-track students have the least-qualified teachers simply because more of them go to disadvantaged, minority schools where the teachers are less-qualified or whether the schools themselves contribute to this matching by systematically assigning their least-qualified teachers to the students they consider the least able.

Even when school-type differences in teacher qualifications are accounted for, most track-level differences remain significant, and others that were obscured by school differences emerge. For example, when community type is controlled for, teachers of low-ability classes have fewer years of experience than teachers of average- and high-ability classes (2 years less, on average).[16] When school SES is accounted for, teachers of classes at lower track levels are less likely to be certified in mathematics and science or to hold bachelor's degrees in these fields—the lowest-ability groups have the fewest qualified teachers, and in all but the highest-SES schools, the highest-ability groups have the most.[17] When racial composition is accounted for, high-track classes have significantly greater exposure to teachers with mathematics or science certification and to teachers with bachelor's degrees than do low-track classes.[18] When community type is accounted for, low-ability classes have less access to teachers with mathematics or science certification and to teachers with either bachelor's or master's degrees in these fields.[19] School differences do not

---

[16]$F = 11.67$, $P < 0.01$.

[17]For certification, $F = 10.40$, $P < 0.01$; for bachelor's degrees, $F = 13.37$, $P < 0.01$. At the highest-SES schools, we found no significant differences in ability groups' access to teachers with bachelor's degrees.

[18]For teacher certification, $P < 0.05$; for teachers' degrees, $P < 0.01$. Average classes did not differ significantly from low-track classes on teacher certification, but they were more like high-ability groups in their access to teachers with degrees.

[19]For certification, $F = 6.18$, $P < 0.01$; for bachelor's degrees, $F = 5.95$, $P < 0.01$; and for master's degrees, $F = 8.79$, $P < 0.01$. In suburban schools, low- and average-track classes had fairly equal access to teachers with bachelor's degrees in science and mathematics, whereas high-ability classes were more likely to have teachers with these degrees.

explain the differences in computer training of teachers of different ability-level classes—again, teachers of high-ability classes are the best qualified in this area.[20] Clearly, secondary schools of all types systematically allocate their most-qualified teachers in ways that disadvantage the students who are thought to be less able in science and mathematics.

While nearly all types of schools place their least-qualified teachers in low-ability classes and their most-qualified teachers in high-ability classes, schools of different types cannot provide students in comparable tracks with teachers who have comparable qualifications. In schools with less-qualified teacher pools (low-SES, high-minority, inner-city schools), teachers of low-track classes are less well-qualified than teachers of low-track classes in schools with generally more qualified staffs (higher-SES, white, suburban, and rural schools). Thus, students at the least-advantaged schools more often compete (through their class assignments) for teachers who are certified to teach mathematics and science or who have bachelor's degrees in these fields. We found that in schools with the highest concentration of low-income students, the teacher qualifications at different track levels differed considerably, as shown in Fig. 4.12.

In contrast, schools whose teachers were generally more qualified exhibited the same patterns of differences, but far more classes had access to certified teachers and teachers with bachelor's degrees. The assignment patterns of teachers at these schools are most evident in terms of the subtle or higher-level qualifications—teachers' perceptions of themselves as master teachers, years of teaching experience (which may represent seniority or political clout in the school, as well as a feeling of high competence), and the holding of master's degrees. These differences are vividly illustrated by contrasting teachers' qualifications in classes of similar ability levels in a subset of *different types of* schools (see Table 4.1). The qualifications of teachers of various track levels at high-minority, low-SES, inner-city schools differed substantially from those of teachers at high-wealth, predominantly white, suburban schools.[21]

---

[20]For computer training, $F = 3.09$, $P < 0.05$.

[21]The first group of schools includes those at which at least 30 percent of the students are from unemployed families or families on welfare, those with minority populations exceeding 50 percent, and those located in inner-city or other urban neighborhoods. The second group includes those at which at least 30 percent of the students have parents in professional or managerial occupations, those with white populations exceeding 50 percent, and those located in suburban neighborhoods.

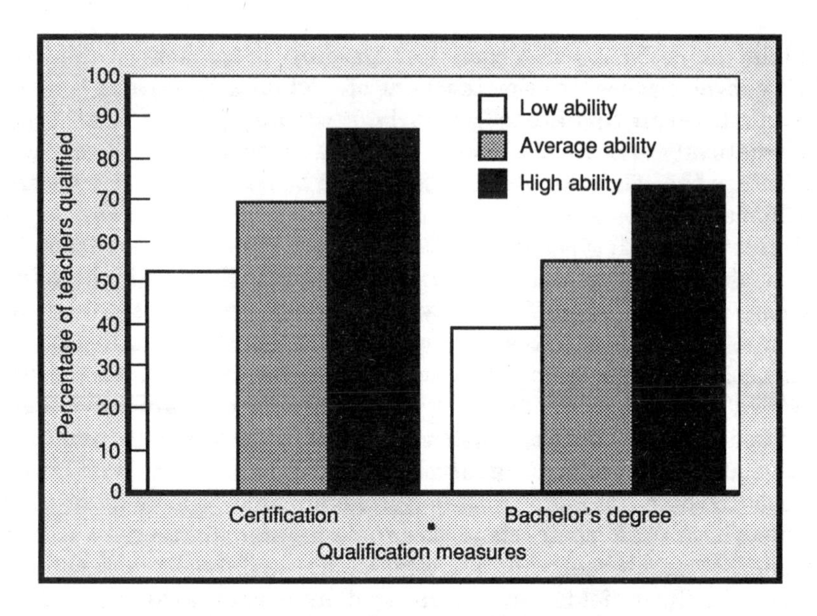

Fig. 4.12—Qualifications of secondary teachers in low-SES schools, by ability level of assigned class

Table 4.1

QUALIFICATIONS OF SECONDARY TEACHERS IN HIGH- AND LOW-ABILITY CLASSES IN SCHOOLS OF DIFFERENT TYPES

	Low-Ability Classes		High-Ability Classes	
Teacher Qualifications	Low-SES, Minority, Urban	High-SES, White, Suburban	Low-SES, Minority, Urban	High-SES, White, Suburban
Certified in science/math	39	82	73	84
Bachelor's in science/math	38	68	46	78
Master's in science/math	8	32	10	48
NSTA qualified	11	36	5	47
NCTM qualified	23	26	4	16
Computer coursework	41	61	69	62

Importantly, however, the differences are somewhat greater for low-track classes than for high-track, especially in terms of teacher certification.

Teachers also differed in their years of teaching experience. Teachers of low-track classes in disadvantaged schools averaged 11.5 years of experience, whereas their counterparts at more-advantaged schools had 15 years of experience. Teachers of high-track classes at disadvantaged schools averaged 13 years, while their peers at advantaged schools averaged 15 years. Interestingly, however, teachers of classes of the same ability level at different types of schools were quite similar in their perceptions of themselves as master teachers: Teachers of high-track classes at both types of schools agreed far more strongly that they were in this category.

Perhaps the most striking finding in this area is that while all types of schools provide high-track students with the best-qualified teachers, the scarcity of highly qualified teachers at disadvantaged schools has enormous implications: *Although high-track students nationwide have access to the best-qualified teachers, low-track students in the most advantaged schools are likely to have better-qualified teachers than high-track students in the least-advantaged schools.* This pattern constitutes a double disadvantage for students judged to be low in ability who attend low-income, predominantly minority schools. Their schools have scarce resources to begin with, and as those schools follow the national pattern of uneven distribution of resources by ability level, low-track students end up with the least of the least. Moreover, their schoolmates who have been judged to be promising fare little better. The "advantages" that accrue to them as the result of being judged able do not even equal those of students judged to be least able and provided the fewest teacher resources at more-advantaged schools.

V. ACCESS TO RESOURCES

This section considers the distribution of resources and materials among different types of schools: computers and special staff to coordinate their instructional use; science laboratories and other science-related equipment and materials; and textbooks. It also examines principals' and teachers' perceptions of whether inadequate resources pose problems for science and mathematics instruction. While there is little evidence that the actual quantity of resources available to schools, teachers, and students has a *direct effect* on learning or willingness to persist in science and mathematics, resources are enablers. They provide the context in which schools and classrooms operate; they often define the outer limits of what is possible. For example, if a school has no science laboratory facilities, even the best-prepared teachers will be unable to engage students in laboratory work.

There are two dimensions of resource availability that seem likely to affect the quality of the instructional program in science and mathematics: the number and type of resources that are available, and the perception of educators about whether or not they have sufficient resources for carrying out their instructional programs. Countable science and mathematics resources would include specialist teachers or coordinators for overseeing programs or providing additional teaching; laboratory facilities; computers and calculators; textbooks; and specialized equipment such as greenhouses, darkrooms, or weather stations.

By counting the number of resources available at different types of schools and comparing those counts across schools, we can determine how evenly some common resources are allocated. Then, by comparing perceptions of how resource adequacy varies across schools, we can identify real inequalities that a simple count of equipment and materials may miss and can show whether schools of different types feel more or less constrained by resources. Different types of schools may perceive their resource needs quite differently, but educators' perceptions of resource adequacy should be taken seriously as indicators of the degree to which resource problems constrain instructional programs.

WHAT SCIENCE AND MATHEMATICS RESOURCES ARE AVAILABLE?

Recent work has documented inequities in the numbers of microcomputers available for student use at different schools and in the ways computers are used for different subpopulations of children (Becker, 1983, 1986; Furr and Davis, 1984; Winkler et al., 1984). In 1986, only about 40 percent of middle schools in low-SES communities had as many as 15 microcomputers, whereas two-thirds of the middle schools in high-SES communities had at least this number (Becker, 1986). The fewest microcomputers were available in elementary schools serving predominantly poor and/or minority children; and at these schools, smaller percentages of children actually used the computers. Moreover, fewer poor and minority schools had teachers who were computer specialists.

The NSSME data permit a broader look at the availability of science and mathematics resources at different types of schools.[1] The data indicate that elementary schools of different types are more similar in terms of availability of science and mathematics resources than are secondary schools.

Elementary Schools

The most obvious finding is that few elementary schools have substantial science and mathematics resources. The nation invests little in elementary mathematics, science, or computer instruction. Overall, there is considerable equality in the *lack* of resources, but elementary schools with large percentages of minority students (i.e., more than 50 percent) are even less likely to have computers available *for student use* than are schools with majority white populations. However, at the time of the NSSME, most schools of both types did have computers—81 percent of the high-minority schools and 94 percent of

---

[1]Principals in the NSSME sample indicated whether their schools had microcomputers; terminals connected to mini/mainframe computers; a greenhouse; a telescope; a darkroom; a weather station; hand-held calculators; microscopes; cameras; scientific models; a small-group meeting room; a learning resource center; mathematics and science laboratories; an outdoor study area; a vivarium; a portable planetarium; a videocassette recorder; and a videodisc player. They also reported the number of computer terminals and microcomputers available for student use and whether anyone on their staff was specifically designated to coordinate or supervise mathematics, science, and computer instruction. Teachers were asked about the availability of computers for use in their classes.

the predominantly white schools.[2] There were no significant differences in the *number of computers* relative to the size of the student body among schools of different socioeconomic or racial composition, but locale did make a difference: Rural elementary schools had more computers available per student than did other schools.[3] This may be attributable to the fact that rural schools typically have fewer students, but they need the same number of computers to set up a computer laboratory that can serve a classroom of children.

As shown in Fig. 5.1, high-SES schools were far more likely than other schools to provide a computer coordinator, and high-minority schools were far less likely to have such a staff person.[4] Urban schools not located in inner cities were also more likely than other

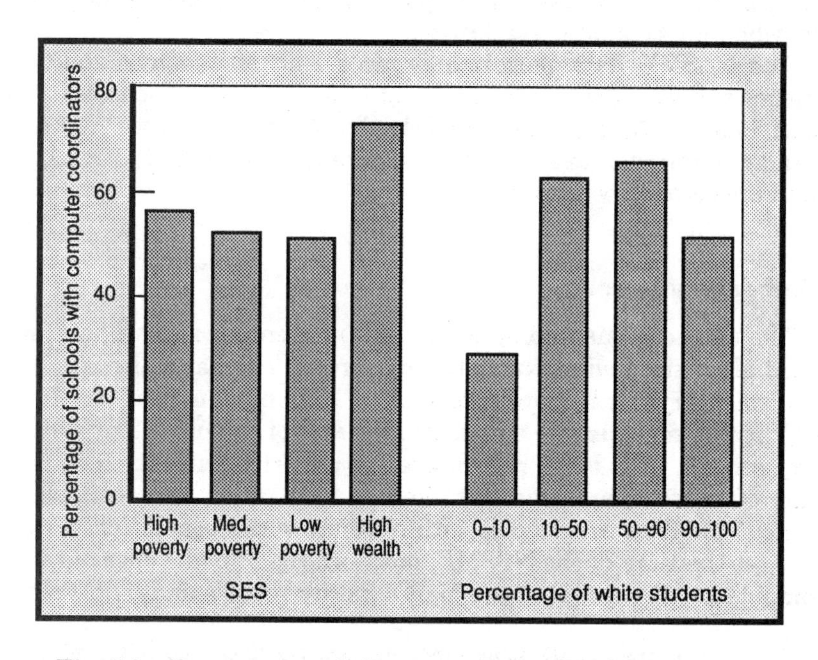

Fig. 5.1—Percentages of elementary schools with computer coordinators, by school SES and racial composition (for SES, $F = 3.48$, $P < 0.05$; for percent white, $F = 4.18$, $P < 0.01$)

---

[2] $F = 3.51$, $P < 0.05$.

[3] Rural schools, on average, had 4.88 computers per 100 students, while inner-city, other urban, and suburban schools averaged about 3 computers per 100 students ($F = 9.87$, $P < 0.01$).

[4] These analyses were performed controlling for school size.

schools to have computer coordinators (70 percent, compared with 53 percent of inner-city schools, 56 percent of suburban schools, and 47 percent of rural schools).[5]

Overall, the availability of a staff coordinator for mathematics and science instruction did not differ at schools of different social class or racial composition (slightly more than a third of all elementary schools had such coordinators). However, urban schools, both inner-city and other urban, were more likely than either suburban or rural schools to have mathematics coordinators—49 percent, compared with 37 and 22 percent.[6]

Both location and racial composition relate to the availability of science facilities, equipment, and materials. However, racial composition stands out in this respect. Elementary schools with a majority of African-American and Hispanic students reported having, on average, only two different types of science-related resources, while schools with majority white populations reported having three.[7] Inner-city schools had fewer materials than schools in other locations, but other urban, suburban, and rural schools had similar resources.[8] While many elementary schools have no science laboratories, some types of schools are more likely to have them than others. Figure 5.2 shows that predominantly white schools are about twice as likely to have science laboratories as predominantly minority schools.

Secondary Schools

Larger differences in the distribution of science and mathematics resources exist at the secondary level.[9] Only 77 percent of the principals of low-SES high schools said that they had computers available for instructional use, whereas 95 percent of the principals of schools in higher SES categories did.[10] Moreover, teachers at low-SES and inner-city schools reported that computers were less readily available at their schools, or, if they were available, they were difficult to secure for use in instruction.[11] Students at high-minority, inner-city

[5]$F = 3.03$, $P < 0.05$.

[6]$F = 6.22$, $P < 0.01$.

[7]$F = 6.70$, $P < 0.01$.

[8]There were an average of 2.4 resources in inner-city schools, 2.6 in other urban schools, 2.8 in suburban schools, and 2.7 in rural schools ($F = 3.36$, $P < .05$).

[9]Each of the analyses of the distribution of resources controlled for the size of schools' student population.

[10]$F = 12.48$, $P < 0.01$.

[11]For school SES, $F = 5.22$, $P < 0.01$; for school location, $F = 6.56$, $P < 0.01$.

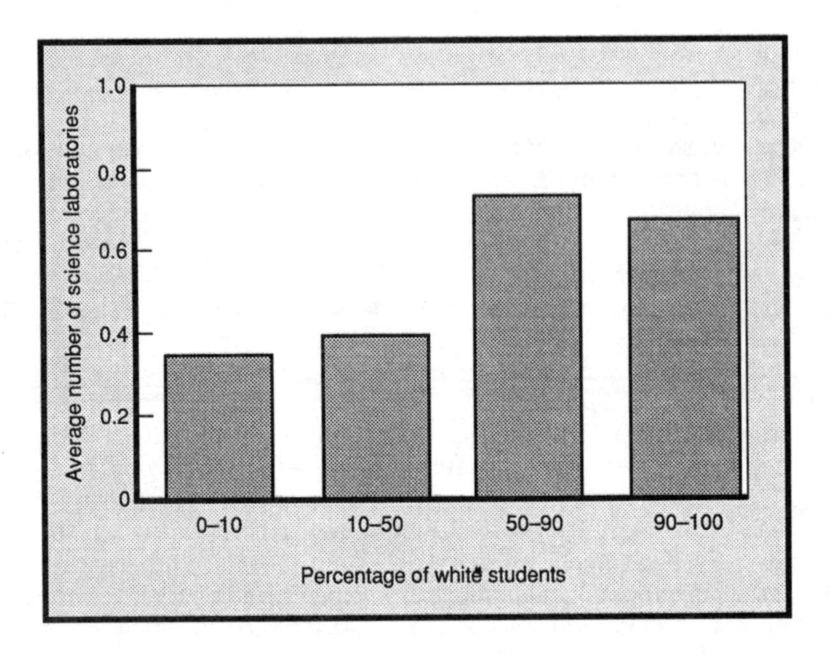

Fig. 5.2—Availability of science laboratories in elementary schools, by school racial composition ($F = 3.13$, $P < 0.05$)

schools had access to *far fewer computers,* even when computers were available. Schools with 90 percent or greater minority populations had an average of 1.76 computers per 100 students; schools with less than 90 percent minorities averaged about 2.70.[12] Inner-city schools averaged 1.88; other urban schools, 2.48; suburban schools, 2.80; and rural schools, 2.86.[13] And, as Fig. 5.3 illustrates, both SES and location affected whether schools were likely to have a staff person designated to supervise or coordinate the instructional use of computers.

Secondary schools also differed in the extent to which they provided a special staff person to supervise or coordinate science and mathematics instruction, with the greatest differences occurring in science (see Table 5.1).

High-SES schools were most likely to have specially designated coordinators for science and mathematics programs. While predomi-

[12] $F = 4.96$, $P < 0.01$.
[13] $F = 8.42$, $P < 0.01$.

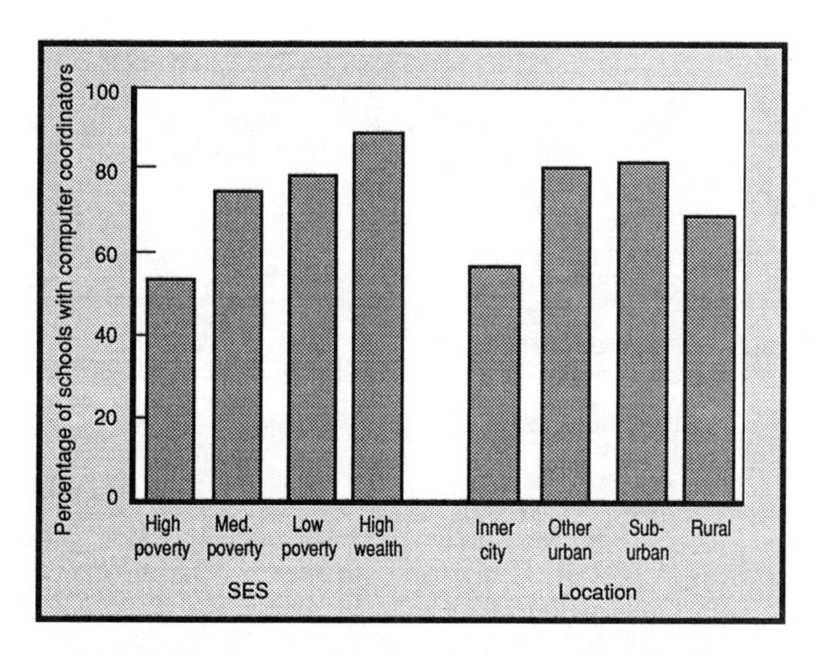

Fig. 5.3—Percentages of secondary schools with computer coordinators, by school SES and racial composition (for SES, $F = 13.43$, $P < 0.001$; for percent white, $F = 6.59$, $P < 0.001$)

nantly white schools were shown to be less likely than minority schools to have such coordinators, the nonwealthy rural schools probably account for this finding. Other interesting differences appear among all types of schools, but wealth and locale seem to be the major factors determining the availability of these human resources.

Secondary schools with large concentrations of low-income or African-American and Hispanic students and schools located in inner cities have far fewer facilities and equipment available, e.g., greenhouses, telescopes, darkrooms, weather stations, calculators, microscopes, cameras, scientific models, outdoor study areas, resource centers, vivariums, or planetariums. An average of 2.33 different types of science equipment were available at the lowest-SES schools, compared with 3.77 at the highest-wealth schools.[14] Schools with the highest concentration of minority students averaged 2.52, and schools

[14] $F = 3.93$, $P < 0.01$.

Table 5.1

PERCENTAGES OF SECONDARY SCHOOLS PROVIDING MATHEMATICS AND SCIENCE COORDINATORS, BY SCHOOL SES, RACIAL COMPOSITION, AND LOCALE

School Characteristic	Mathematics Coordinators	Science Coordinators
SES[a]		
High poverty	67	59
Medium poverty	61	60
Low poverty	64	63
High wealth	93	91
Racial composition[b]		
0–10% white	77	77
10–50% white	83	81
51–90% white	73	67
90–100% white	63	63
Location[c]		
Inner city	74	55
Other urban	75	74
Suburban	76	77
Rural	54	53

[a]Significance of SES: for mathematics, $F = 13.13$, $P < 0.001$; for science, $F = 13.38$, $P < 0.001$.
[b]Significance of racial composition: for mathematics, $F = 4.21$, $P < 0.01$; for science, $F = 2.69$, $P < 0.05$.
[c]Significance of location: for mathematics, $F = 9.60$, $P < 0.001$; for science, $F = 11.59$, $P < 0.001$.

with the lowest, 3.72.[15] Schools in inner cities averaged 2.83; other urban schools, 3.48; suburban schools, 3.56; and rural schools, 3.79.[16]

As at the elementary level, schools with the fewest minority students had slightly more science laboratories than racially mixed or all-minority schools.[17] Location was a far more important factor in this case than any other school characteristic. Inner-city schools reported an average of less than one laboratory; schools in other urban communities, 1.26; suburban schools, 1.34; and rural schools, 1.40.[18]

[15]$F = 14.77$, $P < 0.01$.
[16]$F = 17.20$, $P < 0.01$.
[17]$F = 2.65$, $P < 0.05$.
[18]$F = 8.09$, $P < 0.01$.

DO RESOURCE PROBLEMS HAMPER INSTRUCTION?

Given the differences in the availability of countable resources, it is not surprising that principals and teachers in schools of different types also differed in their perceptions of the effects of resource inadequacy on science and mathematics instruction. In the NSSME, principals and teachers were asked to indicate whether inadequate facilities, insufficient funds for purchasing equipment and supplies, lack of materials for individualizing instruction, insufficient numbers of textbooks, poor quality of textbooks, and/or inadequate access to computers posed instructional problems.[19] The findings are shown in Tables 5.2 and 5.3.

Principals clearly perceived fewer resource problems than teachers did. Principals of elementary schools of different SES composition and locale reported problems at a similar rate. The differences are as expected but are not statistically significant. But when schools were compared in terms of racial composition, principals differed significantly in their perceptions of problems. Principals of high-minority schools reported problems far more often than principals of majority white schools. Differences in secondary school principals' perceptions were more pronounced and were affected by SES, racial composition, and location.

Teachers reported numerous resource problems at both the elementary and secondary levels at all types of schools. This stands to reason, since they experience the problems first-hand as they attempt to teach. Even so, high-poverty, high-minority, and inner-city schools have greater resource constraints that affect teachers' judgments about the quality of their science and mathematics programs.

Taken together, the patterns in the NSSME data are unmistakable. Students in high-poverty, high-minority, and inner-city schools, more than others, have resource constraints that affect the quality of their science and mathematics programs.

---

[19]The wording of the items for principals was slightly different from that for teachers and may contribute to differences in the percentages reporting that resources were a problem. Principals were asked to "indicate if [each of the factors listed above] is a serious problem in [mathematics and/or science]." Teachers were asked to indicate whether each of the factors was "a serious problem," "somewhat of a problem," or "not a significant problem."

Table 5.2

PERCENTAGES OF PRINCIPALS REPORTING RESOURCE PROBLEMS, BY SCHOOL SES, RACIAL COMPOSITION, AND LOCALE

School Characteristic	Elementary Schools	Secondary Schools
SES[a]		
High poverty	19	19
Medium poverty	21	13
Low poverty	16	9
High wealth	15	7
Racial composition[b]		
0–10% white	23	18
10–50% white	28	13
51–90% white	19	15
90–100% white	15	10
Location[c]		
Inner city	20	17
Other urban	16	15
Suburban	18	10
Rural	19	12

[a]Differences between elementary schools by SES are not significant. For secondary schools, $F = 12.35$, $P < 0.001$.

[b]Significance of racial composition: for elementary schools, $F = 4.34$, $P < 0.01$; for secondary schools, $F = 4.21$, $P < 0.01$.

[c]Differences between elementary schools by location are not significant; for secondary schools, $F = 4.51$, $P < 0.01$.

Table 5.3

PERCENTAGES OF TEACHERS REPORTING RESOURCE PROBLEMS, BY SCHOOL SES, RACIAL COMPOSITION, AND LOCALE

School Characteristic	Elementary Schools	Secondary Schools
SES[a]		
High poverty	68	61
Medium poverty	65	57
Low poverty	55	54
High wealth	53	46
Racial composition[b]		
0–10% white	86	64
10–50% white	72	64
50–90% white	59	54
90–100% white	55	53
Location[c]		
Inner city	71	74
Other urban	59	52
Suburban	60	53
Rural	59	53

[a]Significance of SES for elementary schools: $F = 4.77$, $P < 0.01$. For secondary schools, $F = 5.45$, $P < 0.01$.

[b]Significance of racial composition for elementary schools: $F = 9.41$, $P < 0.001$; for secondary schools, $F = 2.81$, $P < 0.05$.

[c]Differences for elementary schools by location are not significant. For secondary schools, $F = 9.89$, $P < 0.001$.

HOW GOOD ARE THE TEXTBOOKS?

While teachers at low-SES, high-minority schools feel the overall pinch of inadequate resources more keenly than do teachers at other types of schools, all of them expressed similar judgments about the quality of the textbooks they have available. However, teachers of low-track classes were generally more dissatisfied with their textbooks than were teachers of higher tracks.

In the NSSME, science and mathematics teachers were asked to rate available texts in terms of appropriateness of reading level; interest to students; clarity and organization; helpfulness in developing problem-solving skills; quality of explanations of concepts; inclusion of examples to reinforce concepts or exercises to practice skills; quality of suggestions for activities and assignments; and quality of supplementary materials.[20] Of particular interest were teachers' ratings of the quality of presentation in the text,[21] whether they thought texts helped students develop problem-solving abilities, and their judgments about the texts' suggestions for activities and assignments and the supplementary materials provided. These aspects of textbooks speak to the quality of the texts per se—independent of the particular students with which they are used.

We aggregated teachers' ratings of textbooks at the school level and then compared the responses from different types of schools. We found no significant differences across schools. However, within schools, teachers of classes of different track levels rated their textbooks somewhat differently. At the elementary level, teachers of low-ability classes rated textbooks (except for supplementary materials) significantly lower than did teachers of average- and high-ability classes.[22] At the secondary level, teachers were generally in agreement about the quality of their texts' suggestions for classroom activities, but on other dimensions, those working with different groups of students had divergent opinions. Teachers of low-track classes gave lower overall ratings for quality of presentation of material in their texts than did teachers of high-track classes.[23] However, on helpfulness in developing students' problem-solving skills and quality of the supplementary materials included with the text, teachers of low-track classes rated their textbooks higher than did other teachers.[24] The more highly qualified teachers of high-ability classes were more critical of texts on these dimensions, perhaps because they know more about their subjects.

[20]Teachers rated the textbook they use most often in class by indicating the strength of their agreement or disagreement with various statements (e.g., "Is not very interesting to my students") on a five-point scale.

[21]We combined three of the statements ("Is unclear and disorganized," "Explains concepts clearly," and "Needs more examples to reinforce concepts" (science only) and "Needs more exercises for practice of skills" (mathematics only)) into a single measure of teachers' perception of the quality of the presentation of content. Responses to negatively worded statements were reversed to obtain a positive rating for the texts.

[22]For overall quality of presentation, $F = 9.54$, $P < 0.01$; for development of problem-solving skills, $F = 6.42$, $P < 0.01$; for quality of suggestions for activities, $F = 3.21$, $P < 0.05$.

[23]$F = 4.37$, $P < 0.05$. Secondary teachers of average classes and low-ability classes gave similar ratings; teachers of high-ability classes, however, rated texts higher than either of these two groups.

[24]For development of problem-solving skills, $F = 10.71$, $P < 0.01$ (again, high-ability teachers stand out—this time for their distinctly lower ratings); for the quality of supplementary materials, $F = 3.48$, $P < 0.01$.

In sum, the quality of available texts appears to be similar across both elementary and secondary schools. However, the opportunities to learn afforded by textbooks differ for students in different track levels. Most significant, textbooks for low-ability classes appear to present science and mathematics content less well than the textbooks used in higher-level classes.

SUMMARY

Our examination of the differences in the distribution of instructional resources that support science and mathematics teaching and learning revealed disturbing patterns of unequal opportunities that parallel our findings concerning science and mathematics programs and teachers. Low-income and minority students who are clustered in schools with others like them and those in inner-city schools have less access to computers, staff to coordinate the use of computers in instruction, science laboratories, and other common science-related facilities and equipment than do students in other schools. Additionally, more principals and teachers at low-SES, high-minority schools report that resource problems create problems for science and mathematics instruction. Finally, across all schools, instruction in low-track classes (again, comprising disproportionate numbers of low-income and minority students) appears to be constrained by science and mathematics texts that teachers judge to be of lower quality in most respects.

VI. CLASSROOM OPPORTUNITIES: CURRICULUM GOALS AND INSTRUCTION

Thus far, we have considered the distribution of opportunities that create boundaries around what students can learn in science and mathematics—extensiveness, content, and rigor of school programs; access of students judged to be of different abilities to science and mathematics courses; allocation of well-qualified teachers; and the availability of important enabling instructional resources. In each case, we have found distressing patterns of fewer opportunities for students who typically exhibit patterns of low achievement and minimal participation in science and mathematics—low-income, African-American, Hispanic, and inner-city students. In this section, we step inside classrooms to examine whether schools and classes of different types also differ in the curricular goals teachers set for their students and in the type of instruction they provide and explore the implications of differences for students' learning opportunities.

CURRICULUM GOALS AND EXPECTATIONS

Some case-study research suggests that even when course titles are the same, the curriculum taught in predominantly poor and minority schools is essentially different from that taught in predominantly white middle- and upper-class schools. These differences suggest that advantaged, white children are more likely to be exposed to essential concepts (as opposed to isolated facts) and to be taught that academic knowledge is relevant to their future lives (Anyon, 1981; Carnoy and Levin, 1986; Hanson, in press). For the most part, however, these issues have received little research attention.

In contrast, there is considerable evidence of differences in the opportunities to learn science and mathematics content in different classrooms within the same school: On average, high-ability groups in elementary schools progress further in a school curriculum over the course of the year (Rist, 1973; Hanson and Schultz, 1978; Barr and Dreeben, 1983; Rowan and Miracle, 1983; Gamoran, 1986). While we know of no systematic studies of content differences in ability-grouped science and mathematics instruction at the elementary level, low-ability reading groups have been shown to spend more time on decoding activities, whereas in high-ability groups more emphasis is placed on the meanings of stories (Alpert, 1974; Hiebert, 1983). High-ability-group students do more silent reading and are interrupted less often when reading aloud (Allington, 1980; Eder, 1981). The high-ability-group advantage is presumably cumulative over the years, and as a result, students with a history of placement in high-ability groups cover considerably more material—and distinctively different material—in elementary school.

Differences in pace and quantity of coverage have also been detected at junior and senior high school levels (Ball, 1981; McKnight et al., 1987; Metz, 1978; Page, 1984). McKnight et al. (1987) used data from the SIMS to examine differences in content for eighth graders enrolled in different types of mathematics classes (e.g., remedial, typical, honors, or algebra). Not only did the lower-level courses provide students with access to fewer mathematics topics and skills, students in lower-level classes in the United States had much narrower curriculum opportunities than their counterparts in many other nations (see Kifer, in press). Not surprisingly, the lack of opportunity to learn various topics was reflected in these students' performance on test items.

Low-track classes not only typically offer a limited array of topics and skills, they consistently emphasize *less-demanding* topics and skills, whereas high-track classes typically include more complex material and more difficult thinking and problem-solving tasks (Burgess, 1983, 1984; Hargreaves, 1967; Metz, 1978; Nystrand and Gamoran, 1988; Oakes, 1985; Powell, Farrar, and Cohen, 1985; Sanders, Stone, and LaFollette, 1987).

In an earlier study of 300 junior and senior high school English and mathematics classes, quantitative and qualitative analyses of data from teacher and student questionnaires, teacher interviews, classroom observations, and content analyses of curriculum packages revealed that high-track students were more often presented with traditional academic topics and intellectually challenging skills (Oakes, 1985). Additionally, teachers in high-track classes more often cited having students learn to be competent and autonomous thinkers as among their most important curricular goals. Teachers of low-track classes more often emphasized basic literacy and computation skills and presented topics commonly associated with everyday life and work. Their important curricular goals focused on conformity to rules and expectations.

CURRICULAR EMPHASIS ACROSS SCHOOLS AND CLASSROOMS

The NSSME data provide useful information about the importance teachers place on central goals of science and mathematics education and about how their expectations vary for different groups of students.

Teachers were asked to rate, on a scale from "none" to "very heavy," the emphasis they placed in a particular class on having students achieve the following objectives:

- Become interested in science/mathematics.
- Learn basic science concepts (science only).
- Know mathematical facts, principles, algorithms, or procedures (mathematics only).
- Prepare for further study in science/mathematics.
- Develop inquiry skills.
- Develop a systematic approach to solving problems.
- Learn to communicate ideas in science/mathematics effectively.
- Become aware of the importance of science/mathematics in daily life.
- Learn about the applications of science/mathematics in technology.
- Learn about the career relevance of science/mathematics.
- Learn about the history of science/mathematics.
- Develop awareness of safety issues in the lab (science only).
- Develop skill in laboratory techniques.

Because both school and classroom characteristics can affect students' access to science and mathematics courses, it is important to understand the emphasis teachers place on various curricular objectives both at schools of different types and in classes of different track levels. Then, to evaluate the relative influence of the school and the classes a student is enrolled in within the school, we must compare the emphasis in classes of the same ability levels in different types of schools.

Elementary Schools. At the elementary school level, about the only differences we found were in the emphasis teachers placed on developing awareness of safety issues in the science laboratory. Teachers in inner-city and rural schools reported emphasizing laboratory safety more than teachers in other urban and suburban settings.

There were no school-level differences related to the racial or socio-economic makeup of the school population.[1]

However, across the sample of elementary schools, we found considerable differences in teachers' emphasis on various objectives in classes that differed in ability level. Table 6.1 shows the strength and direction of these differences.[2]

Table 6.1

ELEMENTARY TEACHERS' CURRICULAR OBJECTIVES: RELATIONSHIP TO CLASS ABILITY LEVEL

Objectives Showing No Significant Positive Relationship with Class Ability Level	Objectives Showing Significant Positive Relationship with High Class Ability Level
Math, facts and principles	Interest *
Math, computations	Science, basic concepts**
Importance in life	Preparation for further study**
Technology applications	Inquiry skills**
History	Problem-solving approach**
Career relevance	Communicate ideas**
Science, lab safety	
Science, lab technique	

NOTE: * = significant at 0.05 level; ** = significant at 0.01 level.

In many respects, teachers have considerably higher expectations for students in high-ability classes. They clearly place more emphasis on some goals that have been widely heralded as critical, not only for future scientists, but for scientifically literate citizens and productive workers in an increasingly technological economy. Such goals as interest in science and mathematics, inquiry skills, and problem-solving are believed to promote essential adult knowledge and competencies; indeed, many science educators suggest that they constitute the core

[1]Where curriculum objectives (or other classroom-level-dependent variables) were analyzed with respect to school characteristics, teachers' responses were averaged within each school and the class weights were summed. In these cases, the number of observations was equal to the number of schools, not the number of teachers surveyed.

[2]The analysis of track level applies only to classrooms with homogeneous grouping; mixed-ability classes were omitted. Class weights were used to provide nationally representative information about differences among classes at different ability levels. Within each category of the respective independent variables, we calculated the teachers' mean response, but because of the difficulty of interpreting the questionnaire's Likert scale responses in absolute terms, we focused primarily on the relative differences in emphasis between categories.

of science and mathematics education.[3] Moreover, teachers place greater emphasis on preparing high-track students for further study in science and mathematics—a goal that might be seen as equally important for the low-track students who are at risk for continuing low achievement and nonparticipation in science and mathematics courses in later grades.

Compounding this unequal access to some important curricular goals, students in low-ability classes are not receiving correspondingly greater emphasis on other curriculum objectives. Teachers of low-ability classes simply seem to set their sights lower than teachers of classes at other track levels.

Secondary Schools. At the secondary level, there are both school and classroom differences in the emphasis teachers place on various objectives. Teachers at high-SES schools emphasize preparing students for further study in mathematics and science, developing inquiry skills and laboratory skills, and acquiring a systematic approach to solving problems.[4] Teachers at lower-SES schools emphasize becoming aware of the importance of science and mathematics in daily life and recognizing the career relevance of these subjects.[5]

We found racial composition to have relatively little effect on teachers' curriculum objectives. At predominantly white schools, teachers place more emphasis on learning basic science concepts; at predominantly minority schools, they place more emphasis on becoming aware of the importance of science and mathematics in daily life.[6]

There are far more differences among classes than among schools. As Table 6.2 illustrates, teachers' emphasis on curriculum objectives differs considerably with the ability composition of their classes. Students in low-track or disproportionately minority classes are disadvantaged in the degree to which teachers emphasize *most* curriculum objectives. Teachers of low-track classes were found to give less emphasis to every curriculum objective except becoming aware of the importance of science and mathematics and performing computations. As at the elementary level, these differences distance students in low-ability classes from some of the most important goals of science and mathematics. Moreover, there is a certain irony to the greater em-

[3]See, for example, Bybee et al., 1989; Champagne and Hornig, 1987.

[4]For preparing students for further study in mathematics and science, $F = 3.63$, $P < 0.05$; for developing inquiry skills, $F = 4.73$, $P < 0.01$; for laboratory skills, $F = 4.32$, $P < 0.01$; for acquiring a systematic approach to solving problems, $F = 6.45$, $P < 0.01$.

[5]For students becoming aware of the importance of science and mathematics in daily life, $F = 5.97$, $P < 0.01$; for the career relevance of these subjects, $F = 4.48$, $P < 0.01$.

[6]For basic science concepts, $F = 3.11$, $P < 0.05$; for becoming aware of the importance of science and mathematics, $F = 3.07$, $P < .05$.

Table 6.2

SECONDARY TEACHERS' CURRICULAR OBJECTIVES: RELATIONSHIP TO CLASS ABILITY LEVEL

Objectives Showing Significant *Negative* Relationship with High Class Ability Level	Objectives Showing No Significant Monotonic Relationship with Class Ability	Objectives Showing Significant *Positive* Relationship with High Class Ability Level
Importance in daily life**	Career relevance	Interest*
Math, computations**		Science, basic concepts**
		Math, facts and principles**
		Preparation for further study
		Inquiry skills**
		Problem-solving approach**
		Communicate ideas**
		Technology applications**
		History**
		Science, lab safety*
		Science, lab techniques**

NOTE: * = significant at 0.05 level; ** = significant at 0.01 level.

phasis teachers of low-ability classes place on developing an appreciation of the importance of science and mathematics in daily life. While few would question the importance of this goal, teachers behave as if they believe that this is all low-track students can do. One might speculate that teachers of low-ability classes work for student appreciation rather than helping their students become knowledgeable and competent.

These track-level differences reveal important nationwide differences in the types of goals teachers emphasize and their expectations for different groups of students. However, because of the uneven distribution of track levels among different types of schools, it is important to understand whether low-track classes receive different curricular emphases partly because they tend to be at schools that emphasize different objectives. In fact, school differences do not appear to account for the ability-level differences noted above. With the influence of school-SES differences accounted for, ability-group differences in teachers' emphasis on preparing students for further study remain.[7] The same is true for developing inquiry skills,[8] laboratory

[7] $F = 95.47$, $P < 0.01$.
[8] $F = 27.83$, $P < 0.01$.

techniques,[9] and a systematic approach to problem solving.[10] However, there is an interesting interaction between school type and track level on some objectives. At the lowest-SES schools, teachers of low-ability classes placed somewhat greater emphasis on inquiry skills and laboratory techniques than did teachers of average classes. In all cases, however, teachers of these two types of classes placed less importance on these objectives than did teachers of high-ability classes.

School differences did not affect the greater emphasis in low-track classes on appreciating the importance of science and mathematics in daily life, but low-SES schools' greater emphasis on the career relevance of these subjects is responsible for ability-group differences. That is, this objective appeared to receive greater emphasis across the sample of low-track classes because it was given greater weight in low-SES schools, and low-track classes were found in far greater numbers in these schools. Thus, track level alone did not produce these differences. Once the differences among schools of different racial compositions were taken into account, ability-group differences in the emphasis placed on learning basic science concepts disappeared. Low-track classes received greater emphasis because there are disproportionately more of them in high-minority schools that place more emphasis on this objective in all types of classes.

Most striking, however, is the finding that teachers of classes at the *same track levels* in very *different types of schools* appear to place similar emphasis on various curriculum objectives.

As shown in Table 6.3, even when the most widely different school types are compared, the curricular emphases in classes at various ability levels are more alike than they are different.[11]

The similarities are particularly noticeable among low-track science and mathematics classes. On only two curricular objectives

[9] $F = 20.01$, $P < 0.01$.

[10] $F = 19.91$, $P < 0.01$.

[11] For these analyses we used two groups of schools: The first included those with at least 30 percent of the students from families that were unemployed or on welfare, those with minority populations exceeding 50 percent, and those located in inner-city or other urban neighborhoods. The second group included those in which at least 30 percent of the students had parents in professional or managerial occupations, those with white populations exceeding 50 percent, and those located in suburban neighborhoods.

Table 6.3

SECONDARY TEACHERS' CURRICULAR OBJECTIVES IN HIGH- AND LOW-ABILITY CLASSES IN SCHOOLS OF DIFFERENT TYPES

More Emphasis in Disadvantaged Schools than in Advantaged Schools[a]	No Significant Difference	More Emphasis in Advantaged Schools than in Disadvantaged Schools[a]
	Low-Ability Classes	
Inquiry skills*	Interest	
History**	Science, basic concepts	
	Math, facts	
	Preparation for further study	
	Problem-solving approach	
	Communicate ideas	
	Inportance in daily life	
	Technology applications	
	Career relevance	
	Science, lab safety	
	Science, lab technique	
	High-Ability Classes	
Importance in daily life**	Interest	Science, lab safety**
Technology applications**	Science, basic concepts	Science, lab technique**
Career relevance**	Math, facts	
History**	Preparation for further study	
	Inquiry skills	
	Problem-solving approach	
	Communicate ideas	

NOTE:* = significant at 0.05 level; ** = significant at 0.01 level.
[a]Disadvantaged schools are low-SES, inner-city or urban, and 50–100% minority schools; advantaged schools are high-SES, suburban, and 0–50% minority schools.

did teachers of low-ability classes in low-SES, predominantly minority, urban schools deviate from their counterparts in high-SES, predominantly white, suburban schools. The curricular focus in high-ability classes was also quite similar across school types. The teachers of these classes in the widely different schools differed on only one objective rated high by more than half of the science teachers and about half of the mathematics teachers: The goal of having students become aware of the importance of science and mathematics in daily life was considered more important in schools serving disadvantaged, minority students.

While school characteristics do influence the curriculum emphases at secondary schools, considerably greater differences result from the judgments educators make about the abilities of their students and the types of class groupings they form. While the NSSME data do not permit causal inferences, school differences appear to stem largely from the disproportionate number of students at high-SES, white schools who are judged to be able learners and the disproportionate number at low-SES, minority schools who are judged to be less able. However, high- and low-track students are generally thought to need much the same curricular focus regardless of where they go to school. The one exception is the more applied and historical curriculum that is offered to low-track students in low-SES, minority schools—a difference that may result from having a less-qualified staff and fewer instructional resources.

LEARNING APPROACHES AND ACTIVITIES

The types of instructional activities that take place in classrooms are useful indicators of *how* teachers go about engaging students in learning. We know of no prior research that has examined differences in instructional practices at the school level or among ability-grouped science and mathematics classes at the elementary level, although considerable case-study and some survey research has investigated the variation in instructional activities with the track level of secondary school classrooms.

Evidence from both American and British ethnographers indicates that teachers describe their expectations for high- and low-track students' classroom participation in different terms (Hargreaves, 1967; Lacey, 1970; Rosenbaum, 1976; Metz, 1978; Ball, 1981; Schwartz, 1981). Hargreaves, for example, found a high-track blackboard with the sign, "We must always remember to behave as an A class," whereas a teacher of a low-ability-level class remarked, "You just can't afford to trust that lot." Such comments seem to be typical of many schools.[12]

Not surprisingly, these differences parallel differences in teaching practices. High-track teachers report spending more time preparing for class, and they appear to be more enthusiastic and more willing to

[12]We are grateful to Reba Page for reminding us of this study.

push their students to work harder (Rosenbaum, 1976; Metz, 1978; Schwartz, 1981; Oakes, 1985). Instruction in low tracks, on the other hand, has been characterized as oversimplified, repetitive, and fragmented. Observers report that teachers of low-track classes use recitation and worksheets to break topics down into minute bits of information, causing lessons to lack overall coherence (Hargreaves, 1967; Keddie, 1971; Metz, 1978; Oakes, 1985; Page, 1987a). Low-track assignments require more rote memory and less critical thinking than work in high-track classes (Hargreaves, 1967; Oakes, 1985). In high-track classes, teachers sometimes pursue serious understanding of complex themes; in low-track classes, instruction is often limited to basic, surface-level understanding of simplified materials (Keddie, 1971; Oakes, 1985; Page, 1987a, 1987b). Even when ostensibly similar materials are used, low-track classes "caricature" other classes in their abbreviated discussions and simplification of ideas. Page (1987b:21) quoted one teacher as saying, "In this particular ninth grade history class, we're less concerned about history and more concerned about improving your reading skills." Thus, students find the "main idea" of a paragraph about the American Revolution, but they do not discuss the implications of the idea itself.

Using national data, Vanfossen, Jones, and Spade (1987) found that college-track students were more likely than others to describe their teachers as patient, respectful, clear in their presentations, and enjoying their work. In earlier work, we found that the use of time also varied by track: In high-track classes, more time and emphasis were devoted to learning activities, and less to behavior management; high-track students also spent slightly more time on-task and were expected to spend more time on homework (Oakes, 1985). In another study, Gamoran (1987) found that high-ability classes were characterized by more open-ended questions, more higher-order cognitive tasks, and more student control over work.

Thus there is strong and consistent evidence of differences in the implementation of curriculum across tracked classes. Reports of fragmentation and rote tasks in low-track classes indicate a consistent pattern of low-quality instruction. This probably also relates to teacher ability and qualifications. Less-qualified teachers have a more limited instructional repertoire and tend to rely on worksheets more often. However, many of the criticisms that have been leveled at low-track classes have also been listed as concerns for the average American classroom. Not only low-track, but also regular classes are described as lifeless, emotionally flat, having fragmented curricula, and including little critical thinking or cognitive challenge (Goodlad,

1984; Powell, Farrar, and Cohen, 1985). Consequently, these differences must be seen within the context of across-the-board classroom instruction that is not very engaging (Gamoran and Berends, 1987).

The NSSME data provide additional insights into how schools and classes enrolling different groups of students vary in the learning activities they provide. The data include the percentages of teachers who used particular types of activities in their last science and/or mathematics lesson and the percentages of class time teachers say students spend on these activities.[13]

Do Learning Activities Differ Among Elementary Schools and Classes?

Elementary teachers were asked which of the following activities they included in the last science or mathematics lesson they taught:

- Lecture
- Discussion
- Student use of computers
- Student use of hands-on materials
- Students doing seatwork assigned from textbook
- Students completing supplemental worksheets

Those reporting on science were also asked whether the following additional activities were included in their most recent lesson:

- Teacher demonstration
- Students working in small groups

Teachers reporting about mathematics lessons were also asked about the following activities:

[13]Teachers were asked to report the instructional activities that took place during the last science or mathematics lesson they taught. While data about a single lesson cannot provide a full picture of time use and activities in any one class, the weighted data can be aggregated to provide representative descriptions for various types of schools (as defined by SES, racial/ethnic composition, and locale). The data also reveal patterns of time use and learning activities among classes of various types (e.g., ability levels). However, the data are limited in that the types of activities listed (e.g., lecture, seatwork, quiz) are *gross* categories, the specific nature of which may differ considerably from class to class. Therefore, analyses of these data cannot begin to portray the subtle differences in the activities or in the teacher-student interactions that take place during instruction—subtleties that can make a tremendous difference in the quality of the instructional opportunities made available to students.

- Student use of calculators
- Tests or quizzes

Teachers at all types of schools reported using basically similar activities in their lessons. Perhaps the most important difference was that a larger proportion of teachers at high-poverty schools used tests or quizzes in their mathematics lessons.[14] Test use also differed among schools of different racial composition, with predominantly minority schools using tests most often.[15] Elementary schools of different types diverged on only one other instructional activity, discussion. A slightly greater proportion of teachers at high-minority schools said that discussion was a part of their most recent lesson.[16]

In addition to reporting the types of learning activities included in their most recent lesson, teachers also indicated how much time they spent on learning activities, daily routines, interruptions, and other noninstructional activities.[17] Science teachers also estimated the time spent on:

- Teacher lecturing
- Students working with hands-on, manipulative, or laboratory materials
- Students reading about science
- Students taking tests or quizzes
- Other science instructional activities

In contrast, mathematics teachers also estimated the time spent in various types of instructional groupings:

- Teacher working with the entire class as a group (e.g., lecture, test, etc.)
- Teacher working with small groups of students

---

[14]Twenty-three percent, as compared with 17 and 15 percent ($F = 2.90$, $P < 0.05$). Because teacher reports were aggregated at the school level, these percentages represent the average percentages of teachers within schools of each type.

[15]Twenty-five percent of the teachers in high-minority schools and 29 percent of those in schools with between 50 and 90 percent minority populations reported that they used tests or quizzes, compared with 18 and 19 percent, respectively, in majority white and 90 percent or more white schools ($F = 2.94$, $P < 0.05$).

[16]Teachers at high-minority schools included discussion somewhat more often (96 percent) than did teachers at other majority-minority schools (88 percent), majority-white schools (85 percent), or predominantly white schools (90 percent) ($F = 3.23$, $P < 0.05$).

[17]Teachers were asked to report the number of total minutes spent on the last lesson and then divide those minutes among the list of activities.

- Teacher supervising students working on individual activities

Although we found few differences in the total amount of time spent on instruction,[18] science teachers at high-minority schools assigned students only about half as much hands-on and laboratory work as teachers at schools with predominantly white enrollments,[19] and those in predominantly minority schools spent twice the time on tests.[20] Also, mathematics teachers at predominantly minority schools had students spend more time working in small groups than did teachers at majority-white schools.[21] While greater amounts of small-group time may appear to provide students greater opportunities for active, engaged learning interaction, in most cases, small-group work actually decreases the amount of instructional time individual students spend with teachers, since the teacher can work with only one group at a time. Individual students, although grouped, often work alone at seatwork while they are waiting for their group's turn with the teacher. Moreover, these findings may well reflect the slightly smaller percentage of homogeneous ability classes in predominantly minority schools, noted in Section II. The greater percentage of class time spent in small groups in these schools probably represents more within-class ability grouping for mathematics instruction.

School location was not a factor in either the distribution of types of activities or the way time was spent.

We found no differences in the *types* of activities that elementary teachers of high-, average-, and low-track classes included in science and mathematics lessons. However, we did find some small differences in their *allocation of class time;* for example, low-track classes spent the most time in class routines.[22] These differences largely reflect the larger number of low-ability classes in low-SES schools,

---

[18]For example, teachers at high-poverty and low-poverty schools spent slightly more time on routines and other noninstructional activities (12 and 11 percent of lesson time, respectively) than did those at moderate-poverty and high-wealth schools (9 percent each) ($F = 4.24$, $P < 0.01$).

[19]Twenty-four percent at schools with between 50 and 100 percent minority; 30 percent at schools with 50 to 90 percent white students; and 48 percent at schools with more than 90 percent white students ($F = 2.89$, $P < 0.05$).

[20] Twelve percent at each of the predominantly minority school types, compared with 6 percent at each of the two types of majority-white schools ($F = 3.44$, $P < 0.05$).

[21] Twenty-two and 26 percent for schools with minority populations greater than 90 percent and 50 to 90 percent, respectively. This compares with 17 percent in schools with between 50 and 90 percent white students and 19 percent for schools with more than 90 percent white students ($F = 3.25$, $P < 0.05$).

[22]Teachers of elementary low-track classes said they spent slightly more time on classroom routines (11 percent) than did teachers of average- (9 percent) or high-track groups (10 percent) ($F = 2.93$, $P < 0.05$).

where routines generally consume more time. Even when we controlled for school differences, however, we found that students in low- and average-ability science classes spent less time on testing[23] and more time on reading than high-ability groups did.[24]

Low-track mathematics classes spent considerably less time than did average- and high-track groups in whole class instruction[25] and considerably more time working with the teacher in small groups.[26] However, once again, these differences are largely a reflection of school differences, although the greater time low-ability classes spend in small groups is not entirely explained by the greater small-group time spent in high-minority schools, where disproportionate percentages of low-ability classes are found.

In summary, we find that teachers in high-minority elementary schools less often involve students in hands-on or laboratory activities. And students in such schools spend more lesson time on routines, testing, and working in small groups than do students in other types of schools. Together, these findings suggest that students in less-advantaged schools have less access to active, engaging learning activities. Track-level differences suggest additional instructional disadvantages for students in low-track classes at these and other types of schools, who spend less time than their peers in other classes actively engaged with the teacher in science and mathematics lessons. In racially mixed schools, because of the placement of large numbers of minority students in low-ability classes, these class-level differences have a disproportionate effect on the opportunities of minority students.

Do Learning Activities Differ in Secondary Schools and Classes?

There is little school-related variation in the types of activities secondary teachers include in their lessons. Neither the composition of the student body nor the location of the school has a noticeable effect on the strategies teachers use in science and mathematics classes. In

---

[23]Five and 7 percent, compared with 12 percent for high-track classes ($F = 3.79$, $P < 0.05$).

[24]Eighteen percent for low-ability, 22 percent for average-ability, and 13 percent for high-ability groups ($F = 5$, $P < 0.05$).

[25]Thirty-five percent, compared with 45 and 47 percent ($F = 8.04$, $P < 0.01$).

[26]$F = 2.96$, $P < 0.05$.

all types of schools, most teachers lecture and few use computers,[27] and most activities specifically related to science or mathematics classes—such as teacher demonstrations,[28] small-group science activities,[29] or the use of calculators in mathematics—are similar across school types.[30]

However, the differences we do find are telling. For example, the percentage of teachers who ask their students to do seatwork is strikingly higher at schools with large concentrations of low-income students—65 percent, compared with 48 percent of teachers at low-poverty/high-wealth schools.[31] Also, nearly half of the teachers in inner-city schools said that they used worksheets in their last lesson, compared with about a third of the teachers in other types of communities,[32] and nearly twice the percentage of teachers in high-minority schools said that their last lesson included a test.[33] The use of hands-on laboratory activities also differed at schools of different SES levels, and seatwork differed with school location. However, the direction and meaning of these relationships are muddy.[34]

There is also considerable divergence in *how much time students spend on different types of activities* at different types of schools. The higher the minority population at schools, the more time teachers spend on daily routines, interruptions, and noninstructional activities, although the size of these differences is small (ranging from 13 percent at schools with minority populations greater than 90 percent

---

[27]For example, 87 percent of all secondary teachers said they lectured during their last lesson; 85 percent said they included discussion; and only 6 percent reported using computers.

[28]Reported by 44 percent of the science teachers.

[29]Thirty-seven percent.

[30]Twenty-one percent.

[31]$F = 6.98$, $P < 0.01$.

[32]Forty-seven percent of the teachers in inner-city schools, compared with 37 percent of suburban teachers, 35 percent of rural, and 31 percent of other urban ($F = 4.47$, $P < 0.01$).

[33]Thirty-one and 26 percent in high-minority and majority-minority schools, compared with 21 percent and 16 percent in majority-white and nearly all-white schools ($F = 6.20$, $P < 0.01$).

[34]High-poverty and high-wealth schools had the lowest percentages of teachers who said that hands-on or laboratory activities were a part of their most recent lesson (24 and 25 percent, respectively); 33 percent of teachers at moderate-poverty and 28 percent at low-poverty schools reported using such activities ($F = 3.85$, $P < 0.01$). Inner-city and suburban schools had the fewest teachers indicating that their students did seatwork (53 percent at each type of school, compared with 61 and 63 percent, respectively, at rural and other urban schools) ($F = 5.99$, $P < 0.01$). However, the low incidence of seatwork in inner-city schools may be accounted for by the greater use of worksheets, as noted above.

to 11 percent at schools with 90 percent or more white populations).[35] More significant, science teachers in schools with higher concentrations of low-income and minority students have their students spend more time reading than do teachers in other schools. Students in the lowest-SES schools spent 14 percent of their class time reading, while those in the high-wealth schools spent only 4 percent. Consistent with this pattern, students in inner-city schools spent more time reading in science classes than did students in other communities.[36] Additional science time spent on reading may come at the expense of instruction delivered directly by the teacher; teachers at the highest-SES schools spent 43 percent of their time lecturing, while those at the lowest-SES schools spent only 33 percent.[37]

Mathematics teachers in high-poverty and majority-minority schools also have their students spend somewhat more time working alone and less time working with the whole class than do teachers with more-advantaged students. Students in high-poverty schools spent, on average, 53 percent of their class time working with the whole class (e.g., listening to teachers' lectures) and 24 percent working alone; students in high-wealth schools spent 60 and 21 percent of their class time, respectively, in these ways.[38]

Overall, then, while there are more similarities than differences in science and mathematics instruction in various types of schools, the pattern of differences is revealing. Students at higher-income and majority-white schools spend more instructional time on whole-class activities and less time working alone, i.e., reading or doing worksheets, than do those at lower-SES, high-minority schools.

The Links Between Tracking and Classroom Activities

Far more striking than the differences between schools of various types are the differences among tracks within schools. Here, too, the differences in how time is spent are greater than the differences in the types of activities teachers include. But, taken together, the differences reveal quite distinct patterns of students in low-track classes spending more time on routine, less engaging, perhaps even less in-

---

[35] $F = 3.03$, $P < 0.05$.

[36] For reading and SES, $F = 10.61$, $P < 0.01$; for reading and racial composition, $F = 8.52$, $P < 0.01$; for reading and school location, $F = 7.60$, $P < 0.01$.

[37] For lecturing and SES, $F = 3.11$, $P < 0.05$; for lecturing and racial composition, $F = 2.75$, $P < 0.05$.

[38] For SES and individual activities, $F = 3.18$, $P < 0.05$; for SES and whole-class activities, $F = 4.05$, $P < 0.01$; for racial composition, $F = 4.20$, $P < 0.01$.

structional activities. Table 6.4 shows that although teachers of the three class levels include most types of instructional activities at the same rates, the pattern of more isolated, routine activities in low-track classes is clear. Students in these classes are more often given seatwork, worksheets, and tests.

Table 6.5 shows how time is divided in classes of different track levels. Students in high-track science classes are advantaged by spending less class time on routines and reading and more time on hands-on activities and receiving instruction from teachers. Students in high-ability mathematics classes spend more time on whole-group instruction and less time working alone.

Considerable literature suggests that the instructional patterns we have observed reflect an overemphasis on control processes and a concomitant deemphasis on educative processes in lower-track classes. In an earlier study where similar instructional differences were found, teachers spent more time disciplining than teaching in lower-track classes (Oakes, 1985). These classes focused on passive drill and practice with trivial bits of information, whereas the upper-track classes included more imaginative, engaging assignments. Other studies describe a similar balance between education and order in high-, average-, and low-track classes, both in the United States and in other industrialized nations, and at the elementary as well as the secondary school level (e.g., Ball, 1981; Eder, 1981; Goodlad, 1984; Hargreaves, 1967; Page, 1987a; Powell et al., 1985; Schwartz, 1981). These findings, combined with evidence that active learning strategies are most likely to promote student achievement in science and mathematics (Bredderman, 1983), suggest that the instructional patterns observed in the NSSME data restrict opportunities to learn in low-ability classes.

The track-level differences remain, even when we control for instructional differences among different types of schools. With the effect of school location accounted for, low-ability groups were found to do seatwork as a part of their lessons far more often than students in other track levels.[39] With school differences in racial composition accounted for, low- and average-ability classes were more often made to complete worksheets,[40] and more teachers of low-ability classes gave tests and quizzes.[41]

---

[39]An average of 63 percent of the lesson time in low-ability classes was spent on seatwork, compared with 45 percent in high-ability classes ($F = 12.44$, $P < 0.01$).

[40]The contrasts between low- and high-track classes and between average- and high-track classes were both significant at the 0.01 level.

[41]Overall ability-group differences were significant at the 0.05 level, $F = 3.04$.

Table 6.4

PERCENTAGES OF SECONDARY TEACHERS INCLUDING VARIOUS INSTRUCTIONAL ACTIVITIES IN LAST SCIENCE OR MATHEMATICS LESSON, BY CLASS ABILITY LEVEL

	Class Type			Significance of Differences	
Instructional Activity	Low	Average	High	F	P <
All classes					
Lecture	89	85	88	2.25	(not significant)
Discussion	88	86	85	1.45	(not significant)
Seatwork	63	61	52	10.20	0.001
Worksheets	43	37	29	16.11	0.001
Small groups	41	37	40	1.17	(not significant)
Hands-on	23	26	27	1.08	(not significant)
Test or quiz	21	18	16	3.07	0.05
Calculators	13	12	25	29.02	0.001
Computers	8	5	6	2.47	(not significant)
Science classes					
Demonstration	46	39	46	3.36	0.05

Table 6.5

PERCENTAGES OF TIME SPENT ON VARIOUS INSTRUCTIONAL ACTIVITIES IN SECONDARY SCIENCE AND MATHEMATICS LESSONS, BY CLASS ABILITY LEVEL

	Class Type			Significance of Differences	
Instructional Activity	Low	Average	High	F	P <
All classes					
Routine	12	12	10	14.08	0.001
Science classes					
Lecture	36	36	41	4.98	0.01
Hands-on	20	20	26	5.85	0.01
Reading	12	10	5	22.58	0.001
Test or quiz	7	7	6	0.26	(not significant)
Other activities	13	14	12	0.96	(not significant)
Mathematics classes					
Class—lecture, test, etc.	48	55	59	18.71	0.001
Small groups	10	9	10	0.65	(not significant)
Individual	29	25	20	14.57	0.001

Ability-group differences in how class time is spent also remain when school characteristics are controlled. The high-track advantage in the smaller amount of time spent on routines remained,[42] as did the greater exposure to teacher-led instruction in these classes.[43] Clear instructional disadvantages for low-track classes also remained after we accounted for school differences. Students in low-track classes across all school types spent greater amounts of their science class time reading.[44] Similarly, low-ability groups spent more time working alone in mathematics and less time doing whole-class activities.[45]

The combined effect of being in a low-track class in a low-SES, high-minority, inner-city school is that lessons tend to be considerably more passive than those in higher tracks at any school. The contrasts shown in Table 6.6 between the most extreme cases—low-track classes in high-poverty, minority, inner-city schools and high-ability classes in high-wealth, white, suburban schools—are particularly striking.

Do Expectations About Homework Differ Among Schools and Class Types?

Finally, we examined how expectations about homework—the instructional time students spend outside of school—differ among schools and classes of different types. We first considered the percentage of teachers who assign homework as a part of their science and mathematics lessons. Then we compared the amount of time teachers in different settings expect students to spend doing homework.

Among elementary schools, neither the concentration of low-income students nor the location of schools made any difference in whether teachers assigned homework or how much time they expected students to spend on it. However, while about a third of the elementary teachers at mixed-race and all-white schools included homework as a part of their science and mathematics lessons, 54 percent of those in

[42]For location and time on routines, $F = 13.77$, $P < 0.01$; for racial composition and routines, $F = 8.48$, $P < 0.02$.

[43]$F = 6.78$, $P < 0.01$.

[44]For SES and reading, $F = 18.96$, $P < 0.01$; for racial composition and reading, $F = 12.07$, $P < 0.01$; for school location and reading, $F = 18.17$, $P < 0.01$.

[45]For racial composition and working alone, $F = 3.94$, $P < 0.05$; for SES and whole-class activity, $F = 9.25$, $P < 0.01$; for SES and working alone, $F = 8.27$, $P < 0.01$; for SES and class activities, $F = 10.74$, $P < 0.01$.

Table 6.6

PERCENTAGES OF TIME SPENT ON VARIOUS INSTRUCTIONAL ACTIVITIES IN HIGH- AND LOW-ABILITY CLASSES IN SECONDARY SCHOOLS OF DIFFERENT TYPES

	Class and School Types	
Instructional Activity	Low-Ability Classes in Low-SES, Minority, Urban Schools	High-Ability Classes in High-SES, White, Suburban Schools
All classes		
Routine	17	9
Science classes		
Lecture	28	51
Hands-on	20	26
Reading	21	1
Test or quiz	10	4
Mathematics classes		
Class—lecture, test, etc.	48	63
Small groups	7	8
Individual	26	20

schools with predominantly minority enrollments assigned homework.[46] We also found large differences in the time teachers expect students to spend on their homework. While teachers in schools with predominantly white populations expected students to spend about 8 minutes on an average day, teachers in high-minority elementary schools expected students to spend twice that much—16 minutes per day.[47] Within elementary schools, the track level of classes made no difference in whether or not teachers assigned homework, but the class ability level did relate to the amount of homework assigned. Teachers in high-track classes assigned students an average of 14 minutes per day, slightly more than the 13 minutes assigned to low-track classes. Teachers assigned average-track classes somewhat less homework, an average of about 10 minutes.[48]

At the secondary level, school type made no difference in the percentage of teachers assigning homework (63 percent across the sampled teachers) or in the amount of time students were expected to spend on homework (an average of 27 minutes per day). Teachers of

[46] $F = 2.85, P < 0.05$.
[47] $F = 9.13, P < 0.01$.
[48] $F = 8.95, P < 0.01$.

classes at all levels were equally inclined to assign homework, but teachers of high-ability classes assigned considerably more homework than other teachers. High-ability classes were assigned an average of 33 minutes of homework per day; average-ability classes, 26 minutes; and low-ability classes, 24 minutes.[49]

The fact that students in low-track classes were expected to spend less time on their homework than other students points to a fundamental irony found in earlier studies of track-level differences in homework (Oakes, 1985). That is, those students who probably need to spend the most time engaged in learning activities to overcome their current deficiencies in science and mathematics are the ones of whom less out-of-school learning time is expected. In contrast, the students who achieve most easily in these subjects are expected to spend the most time learning at home. Thus, teachers have unequal expectations about homework that are likely to further distance high- and low-track students' learning.

SUMMARY

This section has examined two central dimensions of classroom learning opportunities: the curriculum goals that teachers emphasize and the instructional strategies they use to achieve them. Once again, we find patterns that suggest that disadvantaged, minority, inner-city students have more-limited learning opportunities than their more-advantaged, white peers. In the elementary years, differences in curricular goals and instruction are small, but not unimportant. In high-minority elementary schools, there are some small exceptions to the patterns, including higher teacher expectations concerning the amount of homework assigned to students. We must caution, however, that at the elementary level, our measures of opportunity are few and gross in nature, and more work needs to be done on measuring what goes on at this level. It is also possible that the increased investment in instructional time and homework may be having perverse, unintended effects: The additional time may not in fact impart the types of knowledge that encourage participation at a later stage in students' academic careers. The evidence from recent achievement assessments appears to bear this out. Finally, there may be processes at work in middle schools or in the early high school years that undo gains made at the elementary level. Our data show

[49] $F = 80.09$, $P < 0.01$.

that the differences in curricular and instructional opportunities—as in the areas investigated in earlier sections of this report—grow considerably wider in secondary schools.

Again, we find evidence of a double disadvantage for low-income and minority students, particularly in secondary schools. Teachers in schools serving large proportions of these students place somewhat less emphasis on such essential curriculum goals as developing inquiry and problem-solving skills. Moreover, teachers in low-ability classes (where disproportionate percentages of minority students in mixed schools are found) place less emphasis on nearly the entire range of curricular goals.

Schools with large concentrations of low-income and minority students offer fewer classroom conditions that are likely to promote active engagement in mathematics and science learning—such as opportunities for hands-on activities and time working with the teacher. These differences are also compounded by differences in the experiences of students classified as high-, average-, and low-ability. The latter group are disadvantaged in their access to engaging classroom experiences and in their teachers' expectations for out-of-school learning. Because low-income and minority students are disproportionately assigned to low-ability classes, these track-related differences further disadvantage these groups. Thus, our evidence suggests that unequal access to science and mathematics curriculum goals is further exacerbated by discrepancies in instructional conditions in schools and classrooms. Together, the data reveal striking differences in classroom opportunities.

VII. IMPLICATIONS

This study addresses four specific questions about students' opportunities to learn science and mathematics: What science and mathematics are being taught to which students? How? By whom? And under what conditions? The educational system funnels curriculum, resources, instruction, and teachers to students through the schools they attend and the classrooms in which they sit, and this process results in disturbingly different and unequal opportunities to learn—differences that are clearly related to race, social class, community, and the judgments that schools make about students' abilities. At elementary schools, and even more dramatically at junior and senior high schools, science and mathematics programs, teachers, resources, curricular goals, and instructional activities are allocated in ways that disadvantage low-income students, African-American and Hispanic students, and students in inner cities. Those students whom schools judge to have "low ability" and place together in low-track classes are likewise disadvantaged. While each of these characteristics leads to diminished opportunities, students' background characteristics and schools' use of tracked classes combine in ways that place low-income and minority students doubly at risk. Because of the overlap of race, SES, and placement in low-track classes, minority and low-income students' access to learning opportunities is limited beyond what would be expected from being enrolled in either a disadvantaged school or a low-track class.

A CONTEXT OF DIMINISHED RESOURCES AND LOW EXPECTATIONS

Perhaps we should not be surprised by these findings. Children living in communities with low levels of property wealth and personal income typically attend schools that spend fewer dollars on schooling. These relationships appear to persist even in states where school finance reforms have attempted to equalize schooling resources (Carroll and Park, 1983). In fact, per-pupil expenditures between some neighboring high- and low-wealth districts differ by as much as a factor of three or more (*New York Times,* 1990; Wise and Gendler, 1989).

Unequal funding patterns are particularly relevant to race- and social-class equity concerns, since most minority and poor children attend schools in low-wealth communities or in central cities, where the competing demands for tax dollars are greatest. Also, costs of maintaining inner-city schools may be greater even when funding is equivalent. For example, expenditures for building maintenance, replacing items lost or damaged through vandalism, and providing "basic" supplies such as pencils and paper may be greater in older, inner-city schools. In 1983, 71 percent of the African-Americans and 58 percent of the Hispanics in the United States lived in inner-city areas (American Council on Education, 1983). The *proportion* of minority enrollments in large city school districts has increased dramatically in the past fifteen years (in some cases it has doubled), and current projections suggest that this trend will continue. As a result, the pattern of unequal funding in the nation's schools means that poor and minority children will have progressively less access than their more advantaged counterparts to well-maintained school facilities, highly qualified teachers, small classes, and instructional equipment and materials.

Moreover, poor and minority children have been more negatively affected than others by recent changes in educational funding policies. Changes in the method of distributing federal funds have diminished programs and services for disadvantaged children. The Educational Consolidation and Improvement Act (ECIA) of 1981 lessened the regulation and monitoring of Chapter 1 compensatory funds with respect to both targeting aid for particular populations and ensuring comparable spending in target and nontarget schools. Additionally, by combining the Emergency School Assistance Act program (which was aimed at assisting desegregating school districts) with a number of other programs into enrollment-based block grant funding, the ECIA further reduced funds and programs for urban schools and minority children (Darling-Hammond, 1985).

At the state level, decreased public willingness to provide support for schooling (best exemplified by the "tax revolt" that began with the passage of California's Proposition 13 in 1978) has led to substantially fewer dollars being available for education overall. In many advantaged school districts, community groups have offset some of these reductions by establishing educational foundations to raise additional funds. These, however, are not the districts where most poor and minority children live. Even though some state funding is being increased in conjunction with educational reforms, urban districts remain hard-pressed.

Declining enrollments, particularly in urban schools, have further reduced dollars, since state funding is typically allocated on a per-pupil basis. Pressures on educational budgets have caused many urban districts to cut back on the maintenance of facilities and purchases of textbooks and equipment, and some schools have closed altogether. Under these circumstances, we would expect science and mathematics programs in high-poverty, high-minority, and inner-city schools to be adversely affected.

Our findings also underscore the National Science Board's concern that inequalities may stem from the "failure to recognize and develop talent" and "the erroneous belief that many students lack the ability to learn mathematics and science" (NSB, 1983:13). The distribution of opportunity can be understood not only by looking at students' race and social class characteristics, but also by tracing the links between these characteristics, schools' judgments about students' intellectual ability, and the different educational experiences that follow from these judgments.

Educators have historically concluded that track-related differences in teacher expectations, types of knowledge, learning experiences, and even the assignment of teachers were appropriate, given quite apparent differences in students' readiness for particular science and mathematics curricula and instruction. However, our analyses suggest that these differences may reflect something other than appropriate adjustments for differences in aptitude. Students judged to have low ability may get less because they are thought to need less (they are considered unable to benefit) or deserve less (they are considered unwilling to benefit). But one could also argue that they need more—more able teachers, more instructional resources and supports. Here, we suggest that in their efforts to accommodate differences in ability with different educational experiences, schools actually limit some students' opportunities to learn. And given the disproportionately high proportion of students judged to have low ability in schools serving large concentrations of low-income and/or African-American and Hispanic minority students and the disproportionate assignment of minorities to low-track classes in mixed-race schools, these students experience a double disadvantage.

In addition to attending schools with less extensive and less rigorous science and mathematics programs, less-qualified teachers, fewer resources, and less-engaging classroom environments, low-income and minority students often find themselves in low-track classes that focus on "general" mathematics and science content and provide less access to the topics and curricular objectives that could prepare them for successful participation in academic courses in these subjects. They, more than other students, learn in classrooms where instructional activities appear to be directed toward control rather than educative purposes. They interact with less well-qualified and less-confident teachers. In fact, students in high-track classes at high-poverty and high-minority schools may have fewer opportunities than students in low-track classes at more advantaged schools; for example, they may have less access to highly qualified teachers.

These findings raise complex educational and ethical issues. Many schools serving large concentrations of poor children, non-Asian minorities, or inner-city children lack the political clout to command resources equal to those of other schools (although parity with relatively better-funded schools may be too low a standard, given the dire straits inner city and many suburban schools are in). Teachers often view these schools as less desirable places in which to teach, partly because the ever-present difficulties of teaching are compounded by the economic and social disadvantages that shape the students' lives and partly because such schools are often far from the middle-class teachers' homes. Many cities pay less, and inner-city schools have less desirable physical plants (many of them are literally crumbling) and fewer resources for teaching. These factors are often particularly important in teachers' decisions about where to teach. As a result, schools serving disadvantaged and minority students have far greater difficulty attracting and retaining well-qualified teaching staffs.

Within schools, many educators believe they base their decisions about who teaches what science and mathematics, to whom, how, and under what conditions on egalitarian and educationally sound criteria. Although many realize that some decisions are political, rather than educational, and most educators are unhappy about the hiring of unqualified teachers, the processes and outcomes of curriculum differentiation and ability grouping are complex, subtle, often informal, incremental, and usually well-intentioned. For example, high schools with an uneven teaching staff often decide that students studying traditional college-preparatory mathematics content need teachers with stronger preparation in mathematics than do students struggling to understand fundamental mathematics processes. Of course, this assumption can be challenged. Uncertified or unprepared teachers are also least equipped to diagnose students' learning problems or to design activities that will help overcome them.

Nevertheless, considerable evidence suggests that this differentiation, especially at secondary schools, fails to increase learning generally and has the unfortunate consequence of widening the achievement gaps between students judged to be more and less able. Thus, we find no instrumental value that might justify unequal access to valued science and mathematics curricula, instruction, and teachers.

THREE SCENARIOS FOR RIGHTING INEQUALITIES

The inequalities we have documented are not likely to be either self-correcting or easily changed. As long as high-quality educational opportunities are scarce and strategies for teaching diverse groups of students are largely untested, powerful constituencies of advantaged communities and parents will seek to preserve their educational advantages. Policymakers and educators must therefore seek strategies to ameliorate the inequalities and, at the same time, improve the science and mathematics education provided to all students.

We can describe three scenarios for remedying the uneven distribution of science and mathematics resources and teachers. First, through legislative action and local district decisions, policymakers and educators could redistribute the available resources to schools by shifting equipment, materials, and staff away from schools that now have more to those that have less. Second, policymakers at the federal, state, and local levels could attempt to increase the overall levels of educational resources so that all schools could command sufficient facilities, equipment, materials, and qualified staff to develop and sustain high-quality programs. Third, policymakers and educators could work together to both increase educational resources and frame resource allocation policies so that any new resources would go first to those schools serving economically disadvantaged and minority students.

We believe that the first scenario would be politically disastrous, since powerful constituencies would undoubtedly work to prevent a withdrawal of resources from currently advantaged schools. The second also holds little promise, since even as resources increased, the communities and parents who are now more advantaged would undoubtedly use their political clout to garner the lion's share of those resources for their schools and children, absent any policies regulating their distribution. Furthermore, the notion of *advantage* is a relative, rather than an absolute, designation. Schools with better science and mathematics programs may still have legitimate needs for additional resources.

Only the third approach stands a chance of meeting the dual tests of political acceptability and potential effectiveness. Even at that, the extent to which policymakers and educators could marshal support for the preferential distribution of new resources is doubtful. We therefore believe that such an approach stands the best chance of succeeding if it is linked with other policies aimed at improving science and mathematics education at all schools. We suggest some specific targets for such policies below.

POLICIES FOR EQUALIZING OPPORTUNITY AND IMPROVING SCIENCE AND MATHEMATICS EDUCATION

Focusing on the Importance of All Students' Opportunities

In the past decade, policymakers and educators have come to recognize the effectiveness of federal and state "bully pulpits" for drawing public attention to educational problems. National and state leaders have generally used their pulpits to decry poor educational performance and to call for schools to improve student achievement and demonstrate, through accountability indicators, that they have made efforts to do so. However, attention is increasingly turning to the achievement and participation of low-income, minority, and inner-city students. Policymakers worry that an increasingly technological workplace and an increasing percentage of non-Asian minorities in the population portend critical labor-force shortages. Moreover, these human-capital concerns converge with concerns for social and economic equality. As science- and mathematics-related occupations increase in importance in terms of labor-market opportunity, the typically lower attainments of low-income and minority students will increasingly influence their ability to compete for employment and good wages. These concerns have thus focused a great deal of attention on the educational "bottom line," i.e., achievement outcomes.

Policymakers would thus do well to fuel public concern about science and mathematics *opportunities* as well as *outcomes*. Focusing national concern on better and more evenly distributed learning opportunities could clarify the means by which issues of economic prospects and social and economic justice can be addressed. Strong advocacy for such efforts on the part of the federal and state governments could help to establish a receptive climate for policies and practices aimed at better opportunities and fairer distribution of those opportunities.

Generating Resources and Devising Improvement Strategies

Generating new resources inevitably requires money. Additional funding is needed to upgrade science and mathematics facilities and to provide laboratory and computer equipment. Extra money is also needed to raise teachers' salaries to levels that will make science and mathematics teaching more competitive with private-sector science- and mathematics-related careers, as well as to retain experienced teachers who might be attracted to more lucrative jobs outside of schools. With a strong commitment from the White House and the Congress, Washington could probably convince the public that additional federal monies could profitably be allocated to schools serving the most needy children. One example of such funding emerged from recent deliberations regarding the reauthorization of the Carl D. Perkins Vocational Education Act, which will probably channel resources to schools serving large concentrations of disadvantaged students and enable them to *integrate vocational and academic studies*. Science and mathematics are obvious subjects for such integrated programs. Another current effort is the Kennedy and Pell legislative initiative that addresses teacher shortages and distribution. Other creative avenues for generating new public funds could undoubtedly be found if national and state policy leaders turned growing public concern about outcomes toward the importance of educational opportunities. The federal government and individual states might both supplement compensatory education programs targeted specifically toward science education.

Other sources for increased commitment and new resources include the new alliances between business and public education in many cities and states. As it becomes convinced that good public schools play an important role in local and state economic development, business can be effective by generating enthusiasm for new public funding and providing seed money for developing new avenues to equalize science and mathematics opportunities.

Opportunities to learn can also be increased by upgrading the science and mathematics knowledge of current teachers and improving the training of new entrants to teaching. State departments of education and universities can play a central role by developing new curricula and instructional strategies that can serve diverse groups of students, and by supporting research and development on methods of restructuring school organization and curriculum to promote equitable access to resources, knowledge, and teachers. State department staffs, business leadership, and university faculty can also play an important role in upgrading teachers' skills. Building on successful teacher inservice training models such as the University of California's Bay Area Writing Project, science and mathematics faculty could come together with engineers from the private sector in summer workshops with science, mathematics, and vocational teachers to develop new teaching strategies for improving the learning opportunities of low-income and minority students. Scholarships or stipends for participation might be provided to teachers who teach in disadvantaged schools. These new strategies could then be disseminated through networks of schools serving disadvantaged students. Additionally, universities could incorporate training in equitable administrative and teaching practices into education programs for teachers and administrators. Finally, science and mathematics instruction occurs in the context of whole schools. It is unlikely that opportunities in these two subjects can exceed those available throughout the whole schooling enterprise. High expectations and support for children succeeding in *all* academic areas, parent education, and professional conditions for teaching must accompany specific attention to mathematics and science opportunities (Oakes and Lipton, 1990).

Changing Priorities for Resource Allocation

While strategies for improving science and mathematics teaching and learning should be made available to all schools, policies are needed to allocate new material and staff resources first to schools with the greatest need—those that lag behind in computers, laboratories and materials, and well-qualified teachers. Such policies, like other affirmative-action strategies, must be backed by people with the determination to ward off political opposition to what may be seen as unwarranted preferential treatment. Moreover, educational "payoffs" from increased resources to the most troubled schools will not be immediate. Months, years, and terms of office may elapse before the investment in opportunity produces significant returns.

Such determination often is more easily sustained at the federal level. However, state and local policymakers must also frame farsighted policies, since it is at these levels that most educational resources are generated and allocated. For example, successful new efforts to equalize intradistrict funding could send considerable new resources to the most disadvantaged schools. States might also provide incentives for attracting and retaining highly qualified staff at schools serving large concentrations of disadvantaged and minority students. These incentives might include additional funding, technical support for new program development, and public recognition. Local districts can alter resource allocation and teacher assignment policies in ways that keep resources at the most disadvantaged schools. Teacher assignment policies would probably need to be devised cooperatively with teachers' unions and might include rescinding teachers' transfer privileges based on seniority.

Improving the Use of Resources Within Schools

Policies that will equalize and improve opportunities within schools are far more difficult to frame than policies regulating opportunities between schools. Much of what happens inside schools is based on intangible factors such as expectations, beliefs about individual differences and ways to accommodate them, and educators' preferences about which students they want to teach. However, state and local policies aimed at building staff capacity to work effectively with different groups of students should support efforts to restructure schools so that opportunities are equalized for students in the same school who are judged to differ in ability.

As we have shown, tracking in science and mathematics, particularly in secondary schools, channels very different opportunities to different groups of students. Because the differential opportunities that result from ability grouping are related to students' race and social class, and because there is little evidence to support the educational effectiveness of tracked classes, effective alternatives should be sought. Such alternatives will require the development of new school organizational schemes that support efforts to provide equal classroom opportunities. Such schemes might include flexible staffing patterns, such as teams of teachers sharing responsibility for diverse groups of students and/or staggered working hours so that some teaching staff are available to provide extra instructional time after school for students requiring additional help. Other arrangements could involve more flexible use of resources from categorical programs to enable more effective mainstreaming of students with mild learning handicaps who are now served under Special Education programs or educationally disadvantaged students who are now served under Chapter 1 programs, *and their teachers*.

In addition, if schools hope to make science and mathematics learning opportunities accessible to diverse groups of students, they will have to redesign both curriculum and instruction. Help from state departments of education and universities should be an integral part of this process, for both technical and political reasons. Promising curricula and instructional strategies for heterogeneous groups of students already exist and can form the basis for new development. Knowledge gained from research in education, cultural anthropology, and sociolinguistics can support new approaches that may be especially appropriate for low-income and minority students. Nontraditional instruction can be more effective than conventional techniques for minority children. African-American and Hispanic children tend to succeed better in classrooms featuring cooperative, small learning groups (Au and Jordan, 1981; Cohen and DeAvila, 1983; Slavin and Oickle, 1981; Slavin, 1985) and experience-based instruction (Cohen and DeAvila, 1983). Recent analyses of the effectiveness of activity-based science curricula (e.g., the Elementary Science Study, Science—A Process Approach, and The Science Curriculum Improvement Study) have concluded that while all students profit from such curricula, disadvantaged students make exceptional gains in their understanding of science processes, knowledge of science content, and logical development when these methods are used (Bredderman, 1983). Attention to new curricular findings can help ensure that any move away from ability-grouped classes will be accompanied by higher-quality science and mathematics instruction for all students. Such efforts should increase the skills of disadvantaged students and provide the knowledge that will allow them access to rigorous courses in junior and senior high school.

Monitoring the Distribution of Opportunities and Accountability for Equal Opportunity

Finally, given the difficulty of equalizing educational opportunities and the potential political disincentives, federal, state, and local efforts to reach this goal should be carefully monitored. As long as states view public accountability schemes as mechanisms for encouraging local efforts to increase student outcomes, districts and schools should be held accountable for working toward equalizing opportunities. Data systems should be designed to report indicators of school resources, curriculum, teachers, instructional conditions, and outcomes by student race and SES. These indicators could provide insights into the possibilities for new educational policies to interrupt the patterns of unequal opportunities. Monitoring efforts should be complemented by a hierarchy of financial incentives for developing programs that will equalize opportunity, beginning at the federal level and extending to states, communities, and schools.

The following indicators should be useful for monitoring national progress toward equal educational opportunities and holding schools accountable:

- Key resource indicators, by school type (e.g., schools serving different race and SES student populations): per-pupil expenditures, teacher salaries, pupil/teacher ratios, class sizes.
- Instructional time in science and mathematics at elementary schools of different types.
- Course offerings in science and mathematics at secondary schools of different types.
- Ratios of enrollment in different mathematics and science courses to student groups' representation in the school population. (For example, African-American students may represent x percent of the student population, but only y percent of the enrollment in calculus classes.)
- Science, mathematics, and technology resources available at schools of different types.
- Teacher quality at schools of different types.
- Curriculum in science, mathematics, and technology available to different groups of students (e.g., classes serving different race and SES student populations).
- The instructional processes in science, mathematics, and technology that are available to various groups (e.g., in classrooms serving different race and SES student populations).

The experience of schools and districts across the nation clearly shows that accountability systems are powerful tools. Teachers, administrators, and local communities respond both to the data these systems produce and to the implicit message embedded in the nature of their indicators. But such systems are only as good as their design. Indicators and accountability systems directed at monitoring progress toward equal opportunities to learn science and mathematics should be designed not just to reward and/or punish schools, but to enable policymakers to describe and state problems more clearly; to recognize new problems more quickly; and to obtain clues about promising educational programs. Such systems should provide a direct contribution to policymakers' and educators' thinking about issues of equality and educational opportunities, rather than prescriptions for action.[1]

---

[1]See Oakes (1989a) for a more detailed discussion of the use of educational indicators for monitoring equity in science and mathematics education.

Appendix
CLASSIFICATION OF COURSES

This appendix lists the courses offered at the secondary schools in our sample, by title, and explains the basis on which we categorized them as either general, academic/college preparatory, or advanced academic/college preparatory.

The *general* category includes courses in mathematics and science that focus on content not usually considered necessary or appropriate for preparing students for college: remedial courses, applied or vocationally oriented courses, and courses that are either titled General or take a broad, nonrigorous approach to science and mathematics topics. Examples include basic life science, electricity, general science, computational mathematics, business mathematics, general mathematics, and pre-algebra.

Academic courses are defined as those that offer science and mathematics content that is typically considered preparation for college. At the junior high school level, the availability of such courses may enable students to move more quickly into advanced courses when they reach high school. For example, a student who has an opportunity to take algebra in the eighth grade or geometry in the ninth may be on a fast track in high school mathematics and may be able to complete advanced algebra, trigonometry, and calculus before graduation. Similarly, a student who has the opportunity to take biology in junior high school may be able to take more advanced science courses in high school. High school level *academic* courses are those that make up the standard approved sequence of core courses or electives that satisfy minimum college and university entrance requirements.

Advanced academic courses are those (1) whose titles designate them as being both standard academic subjects *and* designated for advanced, accelerated, honors, or "gifted and talented" students or (2) that go beyond the typical minimum requirements for college entrance (e.g., chemistry II, calculus).

To clarify ambiguous course titles, we consulted other researchers who have recently categorized courses for transcript analyses and officials in state departments of education. If no clear category was apparent, we classified courses conservatively, placing them in the lowest categories they qualified for.

The following lists specify the courses in each category across both elementary and secondary school levels.

SCIENCE

General Courses

General Science 7–9
Life Science
Earth Science
Physical Science
General Science 7
General Science 8
General Science 9
General Science 10–12
Ecology, Environmental Science
Other Science
9th & 8th Grade Science
Life/Physical Science
General Science 9–12
Biology/Physical Science
General Biology
Agriculture
Current Issues
Science, grade 6 or under
Earth/Physical Science
General Science, grade unspecified
Medical Technology
Plant Science
Electronics
Applied Chemistry
Life/Earth Science
Aviation
Life/Earth/Physical Science
General Chemistry
Basic Remedial Science
Basic/Fun Physics

Academic/College-Preparatory Courses

Biology I
Chemistry I
Astronomy
Anatomy
Zoology
Earth/Space/Physical Science—Academic
Anatomy/Physiology
Marine Biology
Geology
Oceanography
Meteorology
Chemistry/Physics II
Human Biology
Research
Botany
Microbiology
Cell Biology
Genetics
Embryology
Other Chemistry
Chemistry/Physics I

Advanced Academic/College-Preparatory Courses

Physics I
Biology II
Chemistry II
Physics II
Physiology
AP Biology
AP Chemistry
AP Science
AP Physics
Chemistry/Physics II

MATHEMATICS

General Courses

Mathematics 7
Mathematics 8
General Mathematics 9
General Mathematics 10–12
Business Mathematics
Consumer Mathematics
Remedial Mathematics
Pre-Algebra/Introduction to Algebra
Other Mathematics
Mathematics 7 & 8
Computer Mathematics
General Mathematics 7–9
Mathematics, grade unspecified
General Mathematics 9–12
Mathematics, grade 6 or under
Technical Mathematics
Applied Mathematics

Academic/College-Preparatory Courses

Algebra I
Algebra II
Geometry
Integrated Mathematics
Sequential Mathematics
Advanced Computer Mathematics

Advanced Academic/College-Preparatory Courses

Accelerated Mathematics 7/8/9
Integrated Sequence Accelerated
Trigonometry
Probability/Statistics
Senior Mathematics (no Calculus)
Advanced Senior Mathematics (some Calculus)
Calculus
AP Calculus

Pre-Calculus
Mathematics Analysis
Advanced Mathematics
Honors/Advanced Algebra I
Integrated Mathematics—Trigonometry/Algebra III
Honors Geometry
Integrated Mathematics—Senior Mathematics/Analysis/Calculus
Calculus/Statistics

REFERENCES

Achievement Council (1985). *Excellence for Whom?* San Francisco: Achievement Council.

Allington, R.L. (1980). "Teacher Interruption Behaviors During Primary-Grade Oral Reading." *Journal of Educational Psychology*, 72, 371–374.

Alpert, J.L. (1974). "Teacher Behavior Across Ability Groups: A Consideration of the Mediation of Pygmalion Effects." *Journal of Educational Psychology*, 66, 348–353.

American Association for the Advancement of Science (1984). *Equity and Excellence: Compatible Goals*. Washington, DC: American Association for the Advancement of Science.

American Council on Education (1983). *Demographic Imperatives: Implications for Educational Policy*. Report on the June 8, 1983, forum, "The Demographics of Changing Ethnic Populations and their Implications for Elementary-Secondary and Postsecondary Educational Policy." Washington, DC: American Council on Education.

Anyon, J. (1981). "Social Class and School Knowledge." *Curriculum Inquiry*, 11, 3–42.

Armstrong, J.M. (1980). *Achievement and Participation in Mathematics: An Overview*. Washington, DC: National Institute of Education.

Au, K.H., and Jordan, C. (1981). "Teaching Reading to Hawaiian Children: Finding a Culturally Appropriate Solution." In *Culture and the Bilingual Classroom: Studies in Classroom Ethnography*, edited by H. Trueba and K.H. Au. Rowley, MA: Newbury House.

Ball, S.J. (1981). *Beachside Comprehensive: A Case-Study of Secondary Schooling*. Cambridge: Cambridge University Press.

Barr, R., and Dreeben, R. (1983). *How Schools Work*. Chicago: University of Chicago Press.

Becker, H.J. (1986). *Computer Survey Newsletter*. Baltimore, MD: Johns Hopkins University Center for the Social Organization of Schools.

Becker, H.J. (1983). *School Uses of Microcomputers: Reports from a National Survey*. Baltimore, MD: Johns Hopkins University Center for the Social Organization of Schools.

Berliner, D.C. (1984). "The Half-Full Glass: A Review of Research on Teaching." In *Using What We Know About Teaching*, edited by P.L. Hosford, 51–77. Alexandria, VA: Association for Supervision and Curriculum Development.

Berliner, D.C. (1979). "Tempus Educare." In *Research on Teaching: Concepts, Findings, and Implications*, edited by P.L. Peterson and H.L. Walberg. Berkeley, CA: McCutcheon Publishing.

Berryman, S.E. (1983). *Who Will Do Science?* New York: The Rockefeller Foundation.

Borg, W.R. (1980). "Time and School Learning." In *Time to Learn*, edited by C. Denham and A. Lieberman. Washington, DC: National Institute of Education.

Braddock, J.H. (1990). *Tracking: Implications for Student Race-Ethnic Subgroups* (Technical Report No. 1). Baltimore, MD: Center for Research on Effective Schooling for Disadvantaged Students.

Bredderman, T. (1983). "Effects of Activity-Based Elementary Science on Student Outcomes: A Quantitative Synthesis." *Review of Educational Research*, 53, 499–518.

Brophy, J.E., and Good, T.L. (1986). "Teacher Behavior and Student Achievement." In *Handbook of Research on Teaching*, edited by M.C. Wittrock, 328–375. New York: MacMillan.

Burgess, R.G. (1984). "It's Not a Proper Subject: It's Just Newsom." In *Defining the Curriculum*, edited by I. Goodson and S. Ball. London: The Falmer Press.

Burgess, R.G. (1983). *Experiencing Comprehensive Education: A Study of Bishop McGregor School*. London: Methuen and Co.

Bybee, R., et al. (1989). *Science and Technology Education for the Elementary Years: Frameworks for Curriculum and Instruction.*

Andover, MA: The National Center for Improving Science Education.

California Commission on the Teaching Profession (1985). *Who Will Teach Our Children?* Sacramento, CA: California Commission on the Teaching Profession.

Carey, N. (1989). "Instruction." In *Indicators for Monitoring Mathematics and Science Education: A Sourcebook*, edited by R.J. Shavelson, L.M. McDonnell, and J. Oakes. Santa Monica, CA: The RAND Corporation.

Carnoy, M., and Levin, H. (1986). *Schooling and Work in the Democratic State*. Stanford, CA: Stanford University Press.

Carroll, S.J., and Park, R.E. (1983). *The Search for Equity in School Finance*. Cambridge, MA: Ballinger Publishing Co.

Casserly, P. (1979). *Factors Related to Young Women's Persistence and Achievement in Mathematics*. Washington, DC: National Institute of Education.

Catterall, J.S. (1986). *On the Social Costs of Dropping Out of School*. Program Report No. 86-SEPI-3. Stanford, CA: Stanford Educational Policy Institute.

Champagne, A., and Hornig, L. (1987). "Practical Applications of Theories About Learning." In *Students and Science Learning*. Volume of *This Year in School Science,* edited by A. Champagne and L. Hornig. Washington, DC: American Association for the Advancement of Science.

Chipman, S.F., and Thomas, V.G. (1984). *The Participation of Women and Minorities in Mathematical, Scientific, and Technical Fields*. New York: Howard University.

Cohen, E.G., and DeAvila, E. (1983). *Learning to Think in Math and Science: Improving Local Education for Minority Children* (Final report of the Walter S. Johnson Foundation). Stanford, CA: School of Education, Stanford University.

College Entrance Examination Board. (1985). *National College Bound Seniors, 1985*. New York: College Entrance Examination Board.

Creswell, J.L. (1980). "A Study of Sex Related Differences in Mathematics: Achievement of Black, Chicano, and Anglo Adolescents." As cited in S.M. Malcom, Y.S. George, and M.L. Matyas, *Summary of Research Studies on Women and Minorities in Science, Mathematics and Technology*. Washington, DC: American Association for the Advancement of Science.

Crosswhite, F.J., Dossey, J.A., Swafford, J.O., McKnight, C.C., and Cooney, T.J. (1985). *Second International Mathematics Study: Summary Report for the United States*. Washington, DC: National Center for Education Statistics.

Darling-Hammond, L. (1985). *Equality and Excellence: The Educational Status of Black Americans*. New York: College Entrance Examination Board.

Darling-Hammond, L. (1987). "Teacher Quality and Equality." In P. Keating and J.I. Goodlad, *Access to Knowledge*. New York: College Entrance Examination Board.

Darling-Hammond, L., and Hudson, L. (1989). "Indicators of Teacher and Teaching Quality." In *Indicators for Monitoring Mathematics and Science Education: Background Papers,* edited by R.J. Shavelson, L. McDonnell, and J. Oakes. Santa Monica, CA: The RAND Corporation.

Dossey, J.A., Mullis, I.V.S., Lindquist, M.M., and Chambers, D.L. (1988). *The Mathematics Report Card. Are We Measuring Up?* Princeton, NJ: Educational Testing Service.

Eccles, J., MacIver, D., and Lange, L. (1986). "Classroom Practices and Motivation to Study Math." Paper presented at the Annual Meeting of the American Educational Research Association, San Francisco, CA.

Eder, D. (1981). "Ability Grouping as a Self-Fulfilling Prophecy: A Microanalysis of Teacher-Student Interaction." *Sociology of Education*, 54, 151–161.

Ekstrom, R.B., Goertz, M.E., and Rock, D. (1988). *Education & American Youth*. Philadelphia, PA: The Falmer Press.

Fennema, E., and Peterson, P. (1986). "Autonomous Learning Behaviors and Classroom Environments." Paper presented at the Annual Meeting of the American Educational Research Association, San Francisco, CA.

Fennema, E., and Sherman, J.A. (1977). "Sex-Related Differences in Mathematics Achievement, Spatial Visualization, and Affective Factors." *American Educational Research Journal*, 4, 51–72.

Furr, J.D., and Davis, T.M. (1984). "Equity Issues and Microcomputers: Are Educators Meeting the Challenge?" *Journal of Educational Equity and Leadership*, 4, 93–97.

Gamoran, A. (1987). "The Stratification of High School Learning Opportunities." *Sociology of Education*, 60, 135–155.

Gamoran, A. (1986). "Instructional and Institutional Effects of Ability Grouping." *Sociology of Education*, 59, 185–198.

Gamoran, A., and Berends, M. (1987). "The Effects of Stratification in Secondary Schools: Synthesis of Survey and Ethnographic Research." *Review of Education Research*, 57, 415–435.

Goodlad, J.I. (1984). *A Place Called School: Prospects for the Future.* New York: McGraw-Hill.

Grant, C.A., and Sleeter, C.E. (1986). "Race, Class, and Gender in Education Research: An Argument for Integrative Analysis." *Review of Educational Research*, 56(2), 195–211.

Guthrie, L.F., and Leventhal, C. (1985). "Opportunities for Scientific Literacy for High School Students." Paper presented at the Annual Meeting of the American Educational Research Association, Chicago.

Hallinan, M.T., and Sorenson, A.B. (1983). "The Formation and Stability of Ability Groups." *American Sociological Review*, 48, 838–851.

Hanson, R.A., and Schultz, R.E. (1978). "A New Look at Schooling Effects from Programmatic Research and Development." In *Making Change Happen?*, edited by D. Mann. New York: Teachers College Press.

Hanson, S. (in press). "The College-Preparatory Curriculum Across Schools: Access to Similar Learning Opportunities?" In *Curriculum Differentiation in U.S. Secondary Schools*, edited by Reba Page and Linda Valli. Albany, NY: State University of New York Press.

Hargreaves, D.H. (1967). *Social Relations in a Secondary School.* London: C. Tinling and Company Ltd.

Hiebert, E.H. (1983). "An Examination of Ability Grouping for Reading Instruction." *Reading Research Quarterly*, 18, 231–255.

Husen, T. (ed.) (1967). *International Study of Achievement in Mathematics: A Comparison of Twelve Countries* (Vols. 1 and 2). New York: John Wiley.

Johnston, K.L., and Aldridge, B.G. (September/October 1984). "The Crisis in Education: What is It? How Can We Respond?" *Journal of California Science Teachers*, 19–28.

Jones, L.V. (1984). "White-Black Achievement Differences: The Narrowing Gap." *American Psychologist*, 39, 1207–1213.

Jones, L.V., Burton, N.W., and Davenport, E.C. (1984). "Monitoring the Achievement of Black Students." *Journal for Research in Mathematics Education*, 15, 154–164.

Jones, L.V., Davenport, E.C., Bryson, A., Bekhuis, T., and Zwick, R. (1986). "Mathematics and Science Test Scores as Related to Courses Taken in High School and Other Factors." *Journal of Educational Measurement*, 23(3), 197–208.

Kagan, S. (1980). "Cooperation-Competition, Culture, and Structure Bias in Classrooms." In *Cooperation in Education*, edited by S. Sharon. Provo, UT: Brigham Young University Press.

Keddie, N. (1971). "Classroom Knowledge." In *Knowledge and Control*, edited by M.F.D. Young. London: Collier-Macmillan.

Kifer, E. (in press). "Opportunities, Talents and Participation." In *Second International Mathematics Study: Student Growth and Classroom Process in the Lower Secondary Schools*, edited by L. Burstein. London: Pergamon Press.

Lacey, C. (1970). *Hightown Grammar: The School as a Social System*. Manchester: Manchester University Press.

Lantz, A.E., and Smith, G.P. (1981). "Factors Influencing the Choice of Nonrequired Mathematics Courses." *Journal of Educational Psychology*, 73, 825–837.

Lee, V.E. (1986). "The Effect of Curriculum Tracking on the Social Distribution of Achievement in Catholic and Public Secondary Schools." Paper presented at the Annual Meeting of the American Educational Research Association, San Francisco, CA.

Lee, V.E., and Bryk, A.S. (1988). "Curriculum Tracking as Mediating the Social Distribution of High School Achievement." *Sociology of Education*, 62, 78–94.

Levin, H.M. (1986). *Educational Reform for Disadvantaged Students: An Emerging Crisis*. Washington, DC: National Education Association.

Levy, G.E. (1970). *Ghetto School*. New York: Pegasus Press.

Lockheed, M.E. (1985). *Understanding Sex/Ethnic Related Differences in Mathematics, Science, and Computer Science for Students in Grades Four to Eight*. Princeton, NJ: Educational Testing Service.

Lockheed, M.E. (1984). "Sex Segregation and Male Preeminence in Elementary Classrooms." In *Women in Education*, edited by E. Fennema and M.J. Ayer. Berkeley, CA: McCutchan.

New York Times (1990). "Courts Ordering Financing Changes in Public Schools," March 11, 1990, Section 1, pp. 1, 13.

Maccoby, E.E., and Jacklin, C.N. (1974). *The Psychology of Sex Differences*. Stanford, CA: Stanford University Press.

McKnight, C., Crosswhite, F.J., Dossey, J.A., Kifer, E., Swafford, J.O., Travers, K.J., and Cooney, T.J. (1987). *The Underachieving Curriculum*. Champaign, IL: Stipes Publishing.

Metz, M.H. (1978). *Classrooms and Corridors: The Crisis of Authority in Desegregated Secondary Schools*. Berkeley, CA: University of California Press.

Mullis, I.V.S., and Jenkins, L.B. (1988). *The Science Report Card: Elements of Risk and Recovery*. Princeton, NJ: Educational Testing Service.

National Alliance of Black School Educators (1984). *Saving the African American Child*. Washington, DC: National Alliance of Black School Educators.

National Center for Education Statistics (1985). *The Condition of Education, 1985 Edition*. Washington, DC: U.S. Department of Education.

National Center for Education Statistics (1983). *The Condition of Education, 1983 Edition*. Washington, DC: U.S. Department of Education.

National Science Board, Commission on Precollege Education in Mathematics, Science, and Technology (1983). *Educating Americans for the 21st Century: Source Materials*. Washington, DC: National Science Foundation.

National Science Board (1987). *Science & Engineering Indicators, 1987*. NSB 87–1. Washington, DC: National Science Board.

National Science Foundation (1988). *Women and Minorities in Science and Engineering*. NSF 88–301. Washington, DC: National Science Foundation.

National Science Teachers Association (1987). *Survey Analysis of U.S. Public and Private High Schools: 1985–1986* (Draft). Washington, DC: National Science Teachers Association.

Nystrand, M., and Gamoran, A. (1988). *A Study of Instruction as Discourse*. Madison, WI: Wisconsin Center for Education Research.

Oakes, J. (1990). *Lost Talent: The Underparticipation of Women, Minorities, and Disabled Students in Science*. Santa Monica, CA: The RAND Corporation.

Oakes, J. (1989a). "Creating Indicators that Address Policy Problems: The Distribution of Opportunities and Outcomes." In *Indicators for Monitoring Mathematics and Science Education: A Sourcebook*, edited by R.J. Shavelson, L.M. McDonnell, and J. Oakes. Santa Monica, CA: The RAND Corporation.

Oakes, J. (1989b). "School Context and Organization." In *Indicators for Monitoring Mathematics and Science Education: A Sourcebook*, edited by R.J. Shavelson, L.M. McDonnell, and J. Oakes. Santa Monica, CA: The RAND Corporation.

Oakes, J. (1987). "Tracking in Secondary Schools: A Contextual Perspective." *Educational Psychologist*, 22, 129–154.

Oakes, J. (1985). *Keeping Track: How Schools Structure Inequality*. New Haven, CT: Yale University Press.

Oakes, J. (1983). "Limiting Opportunity: Student Race and Curricular Differences in Secondary Vocational Education." *American Journal of Education*, 91, 801–820.

Oakes, J., Gamoran, A., and Page, R. (in press). "Curriculum Differentiation." In *Handbook for Research on Curriculum*, edited by P.W. Jackson. New York: MacMillan.

Oakes, J., and Lipton, M. (1990). Making the Best of Schools: A Handbook for Parents, Teachers, and Policymakers. New Haven: Yale University Press.

Orland, M.E. (1988). "The Demographics of Disadvantage: Intensity of Childhood Poverty and Its Relationship to Educational Achievement." In *Access to Knowledge*, edited by J. Goodlad and P. Keating. New York: College Entrance Examination Board.

Page, R. (1987a). "Lower-Track Classes at a College-Preparatory School: A Caricature of Educational Encounters." In *Interpretive Ethnography of Education: At Home and Abroad*, edited by G. Spindler and L. Spindler. Hillsdale, NJ: Erlbaum.

Page, R. (1987b). "Teachers' Perceptions of Students: A Link Between Classrooms, School Cultures, and the Social Order." *Anthropology and Education Quarterly*, 18, 77–99.

Page, R. (1984). "Perspectives and Processes: The Negotiation of Educational Meanings in High School Classes for Academically Unsuccessful Students." Ph.D. dissertation, University of Wisconsin-Madison.

Pallas, A.M., and Alexander, K.L. (1983). "Sex Differences in Quantitative SAT Performance: New Evidence on the Differential Coursework Hypothesis." *American Educational Research Journal*, 20, 165–182.

Parsons, J.E., Kaczala, C.M., and Meece, J.L. (1982). "Socialization of Achievement Attitudes and Beliefs: Classroom Influences." *Child Development*, 52, 322–339.

Pascal, A. (1987) *The Qualifications of Teachers in American High Schools*. Santa Monica, CA: The RAND Corporation.

Peng, S.S., Owings, J.A., and Fetters, W.B. (1981). "Effective High Schools: What Are Their Attributes?" Paper presented at the 1981 Annual Meeting of the American Statistical Association, Cincinnati, OH.

Persell, C.H. (1977). *Education and Inequality: The Roots and Results of Stratification in America's Schools*. New York: The Free Press.

Peterson, L., and Fennema, E. (1985). "Effective Teaching, Student Engagement in Classroom Activities, and Sex-Related Differences in Learning Mathematics." *American Educational Research Journal*, 22, 309–335.

Powell, A., Farrar, E., and Cohen, D.K. (1985). *The Shopping Mall High School*. Boston: Houghton-Mifflin.

Rist, R. (1973). *The Urban School: Factory for Failure*. Cambridge, MA: MIT Press.

Rock, D., Braun, H.I., and Rosenbaum, P.R. (1985). *Excellence in High School Education: Cross-sectional Study, 1980–1982. Final Report*. Princeton, NJ: Educational Testing Service.

Rock, D., Braun, H.I., and Rosenbaum, P.R. (1984). *Excellence in High School Education: Cross-sectional Study, 1972–1980. Final Report*. Princeton, NJ: Educational Testing Service.

Rosenbaum, J.E. (1986). "Institutional Career Structures and the Social Construction of Ability." In *Handbook of Theory and Research for the Sociology of Education*, edited by J.G. Richardson. New York: Greenwood Press.

Rosenbaum, J.E. (1980). "Social Implications of Educational Grouping." *Review of Research in Education*, 8, 361–401.

Rosenbaum, J.E. (1976). *Making Inequality: The Hidden Curriculum of High School Tracking*. New York: John Wiley and Sons.

Rosenholtz, S., and Simpson, C. (1984). "The Formation of Ability Conceptions: Developmental Trend or Social Construction." *Review of Educational Research*, 54, 31–63.

Rowan, B., and Miracle, A.W., Jr. (1983). "Systems of Ability Grouping and the Stratification of Achievement in Elementary Schools." *Sociology of Education*, 56, 133–144.

Sanders, N., Stone, N., and LaFollette, J. (1987). *The California Curriculum Study: Paths Through High School*. Sacramento, CA: California State Department of Education.

Schmidt, W.H. (1983). "High School Course Taking: A Study in Variation." *Journal of Curriculum Studies*, 15(2), 167–182.

Schwartz, F. (1981). "Supporting or Subverting Learning: Peer Group Patterns in Four Tracked Schools." *Anthropology and Education Quarterly*, 12, 99–121.

Sells, L. (1982). "Leverage for Equal Opportunity Through Mastery of Mathematics." In *Women and Minorities in Science: Strategies for Increasing Participation,* edited by S.M. Humphreys. Washington, DC: American Association for the Advancement of Science.

Slavin, R.E. (1990). *Tracking and Achievement in Secondary Schools: A Review of Research.* Madison, WI: National Center for Research on Effective Secondary Schools, University of Wisconsin.

Slavin, R.E. (1987). *A Review of Research on Elementary Ability Grouping.* Baltimore, MD: Johns Hopkins University Press.

Slavin, R.E. (1985). "Cooperative Learning: Applying Contact Theory in Desegregated Schools." *Journal of Social Issues,* 41, 45–62.

Slavin, R.E. (1983). *Cooperative Learning.* New York: Longman.

Slavin, R.E., Braddock, J.H., Hall, C., and Petza, R.J. (1989). *Alternatives to Ability Grouping.* Baltimore, MD: Center for Research on Effective Schooling for Disadvantaged Students.

Slavin, R.E., and Oickle, E. (1981). "Effects of Cooperative Teams on Student Achievement and Race Relations." *Sociology of Education,* 55, 174–180.

Task Force on Women, Minorities, and the Handicapped in Science and Technology. (1988). *Changing America: The New Face of Science and Engineering* (Interim report). Washington, DC: Task Force on Women, Minorities, and the Handicapped in Science and Technology.

Tobin, D., and Fox, L.H. (1980). "Career Interests and Career Education: A Key to Change." In *Women and the Mathematical Mystique,* edited by L.H. Fox and D. Tobin. Baltimore: Johns Hopkins University Press.

Travers, K.J., and Westbury, I. (1989). *The IEA Study of Mathematics I: Analysis of Mathematics Curriculum.* Oxford: Pergamon.

Valli, L. (in press). "A Curriculum of Effort: Tracking Students in a Catholic High School." In *Curriculum Differentiation in U.S. Secondary Schools: Interpretive Studies,* edited by Reba Page and Linda Valli. Albany, NY: State University of New York Press.

Vanfossen, B.E., Jones, J.D., and Spade, J.Z. (1987). "Curriculum Tracking and Status Maintenance." *Sociology of Education*, 60, 104–122.

Vanfossen, B.E., Jones, J.D., and Spade, J.Z. (1985). "Curriculum Tracking: Causes and Consequences." Paper presented at the Annual Meeting of the American Educational Research Association, Chicago.

Weiss, I. (1987). *Report of the 1985–86 National Survey of Science and Mathematics Education*. Research Triangle, NC: Research Triangle Institute.

Welch, W.W., Anderson, R.E., and Harris, L.J. (1982). "The Effects of Schooling on Mathematics Achievement." *American Educational Research Journal*, 19, 145–153.

Winkler, J.D., Shavelson, R.J., Stasz, C., Robyn, A., and Feibel, W. (1984). *How Effective Teachers Use Microcomputers for Instruction*. Santa Monica, CA: The RAND Corporation.

Wise, A.E., and Gendler, T. (1989) "Rich Schools, Poor Schools: The Persistence of Unequal Education," *The College Board Review*, No. 151, Spring, 12–17, 36–37.

Wise, A.E., et al. (1987). *Effective Teacher Selection: From Recruitment to Retention*. Santa Monica, CA: The RAND Corpora-tion, R-3139-NIE/CSTP.

Wolf, R.M. (1977). *Achievement in America*. New York: Teachers College Press.